Elements of Pig Science

C. T. Whittemore

Professor of Animal Production,
Edinburgh School of Agriculture

Longman Scientific & Technical,
Longman Group UK Limited,
Longman House, Burnt Mill, Harlow,
Essex CM20 2JE, England
and Associated Companies throughout the world.

Copublished in the United States with
John Wiley & Sons, Inc., 605 Third Avenue, New York,
NY 10158

First published 1987

British Library Cataloguing in Publication Data

Whittemore, Colin T.
Elements of pig science.——(Longman handbooks in agriculture)
1. Swine
I. Title
636.4 SF395

ISBN 0-582-45599-5

Library of Congress Cataloguing-in-Publication Data

Whittemore, Colin Trengove.
Elements of pig science.

(Longman handbooks in agriculture)
Includes index.
1. Swine. I. Title. II. Title: Pig science.
III. Series.
SF395.W467 1987 636.4 86–21142
ISBN 0–470–20744–2 (USA only)

Produced by Longman Group (FE) Limited
Printed in Hong Kong

Elements of pig science

Longman Handbooks in Agriculture

Series editors

C. T. Whittemore
K. Simpson

Books published

C. T. Whittemore: *Lactation of the dairy cow*

C. T. Whittemore: *Pig production – the scientific and practical principles*
A. W. Speedy: *Sheep production – science into practice*
R. H. F. Hunter: *Reproduction of farm animals*
K. Simpson: *Soil*
J. D. Leaver: *Milk production – science and practice*
J. M. Wilkinson: *Beef production from silage and other conserved forages*
E. Farr and W. C. Henderson: *Land drainage*
K. Simpson: *Fertilizers and manures*

Contents

To Chris, Joanna, Jonathan,
Emma and Rebecca

Preface

The scientific elements of pig production are the basis for a major biological business providing meat for human consumption. The pig industry is at the forefront of the application of science to practical production technologies; and much of its success is due to the willingness and enthusiasm with which new information and novel innovations are commandeered and applied to primary pig production, pig breeding, and to the associated animal feed and human food industries.

Pigs are especially useful subjects for animal research, and it is perhaps for this reason that pig science has so much to offer the biological sciences in general, and the animal sciences in particular. There is no doubt that working with pigs gives a scientist the dual privilege of an excellent experimental animal and an industry ready to use research findings at short notice.

What this book may have to offer to both the science and the practice of animal production is a direct consequence of the accommodating nature of the pig, of the forward-looking approach of pig farmers and those working in the allied industries of feed and pharmaceuticals, and lastly of the small group of colleagues spread about the world who work so closely together to advance the understanding of pig science. To pigkeepers and pig scientists, a debt of gratitude is owed by the author, and by all who enjoy the privilege of eating pork, bacon, ham and that huge variety of pigmeat products which are now so familiar. It is hoped that this book may serve to draw to the attention of scientists a new approach to animal research interpretations and also to enable more efficient and caring methods of pig production. As is apparent, the text has drawn particularly heavily upon the work of colleagues at Edinburgh; but also on the work of others elsewhere in the United Kingdom, especially the National Research Institute for Research in Dairying (now AGRI) at Shinfield, the Rowett Research Institute at Aberdeen, and the University of Nottingham. The author acknowledges his sources of information and inspiration, and is grateful for the scientific companionship that has resulted from collaborative efforts.

Chapter 1 The nature of animal science

There is little sense, and no science, in building complicated information structures on an illogical foundation. Yet often in science, information accumulation takes precedence over consideration of the principles of the scientific elements. Animal science requires more of logic than it does of dogma and more of common sense than misused statistics.

Experiments

Doing experiments is not science; it is only a part of science. The value of science lies more in the selection of the right experiment and the interpretation of its result.

The discipline of statistical analysis can be a powerful force for good in helping interpretation of experimental results and giving guidance concerning which of the phenomena that are observed are likely to be repeatable, and which merely chance. But the same discipline can also become stultifying and stand in place of, rather than as an aid to, interpretation. Relationships, however statistically closely correlated, should not be confused with causes; nor statistical significances, however great, confused with the establishment of facts.

The contemporary scientific literature is full of reports of factorial experiments, subjected to analysis of variance, and reported in terms of treatments being better than or worse than controls. So often, such experimentation masquerades as science, while the net effect has often been obfuscation rather than the advancement of human understanding of the animals under his care.

Classical investigation method requires first a hypothesis – an expectation of the course of events; next an experimental regime to test rigorously the hypothesis at its most sensitive point; and lastly, a preconceived notion of how the results obtained may be satisfactorily analysed and interpreted in a way that will allow the experimental conclusions to be unequivocally conveyed to others – not so that they may learn and agree, but so they might challenge and disagree. It is by dissent and not assent that animal science progresses. Yet frequently, experiments may be set up with no deeper remit than to test whether one treatment is better than another, no more penetrating design than to cover a treatment range and no view whatsoever as to how the experiment may conclude, nor how any of the possible conclusions of the work may be satisfactorily explained to others and backed by the logic of biological, physical or chemical expectations. It is a commonplace, but often forgotten, part of the miracle of sight that there must be a preconception of what is being looked at before it can be seen clearly. While correlations, regressions and significant differences are useful diagnoses for the phenomena of animal life, they should not be confused with, or used for, the identification of causes and the evolution of a prognosis.

In circumstances where ignorance is all-embracing, where there is no means of support for any hypotheses or preconceptions of experimental consequences, where all is mystery and magic, then the purely empirical investigation is the only course open to the research worker. However, in this event the experimenter must be cognisant of the limitations of his work and the restricted nature of its potential applications.

There is, for example, no causal hypothesis to deal with the different proportions of lean, fat and bone in the carcasses of pigs carrying the halothane gene as compared with conventional White strains. Nevertheless, it is important that these differences are quantified, be they understood at this moment in time or not. In the 1930s, experimentation with raw potato as a feed source for pigs demonstrated that slower growth ensued unless the diet was supplemented with extra protein (at that time in the form of fish meal). It was further noted that this was not the case if the potato was properly cooked. This information was vital to the efficient use of the potato in pig diets and allowed pig production to continue through subsequent cereal shortages in many parts of Europe. It was not until a quarter-century later that the potato was realized to be a potent source of trypsin-inhibitor activity. Trypsin inhibitors disrupt protein digestion and, being proteins themselves, are also destroyed by heat. In this case, knowing the cause has certainly advanced scientific understanding, but identifying the phenomenon, even in the absence of a known cause, was of incalculable benefit.

But often the empirical experimental route is chosen because with its naivety comes simplicity and the abrogation of the scientific responsibility of interpretation. Often deductive experimentation could be utilized, but is not. Determination of the energy needs of pigs by empirical examination of a range of feeding levels may risk the erroneous conclusion that all pigs need only the amount of food found to be satisfactory at one or two research centres. Without an understanding of the causal relationships between nutrient use and nutrient supply, there is some risk of empirical conclusions endangering the efficiency of sizeable portions of the industry.

The appropriate philosophy must be that wherever the deductive approach to experimentation can be used, then it should be. It is in the identification of causal forces that lies the means to understanding the way animals work. Finding out *what* goes on, and explaining *why* it goes on, are dual functions that must go hand in hand. Unfortunately, much of the physical paraphernalia of science, the aspirations of scientific management and the means toward scientific commendation, favours the *what* approach and mitigates against the more difficult pursuit of *why*. When the two aspects of science get out of phase, science runs the risk of being discredited because it is so much more liable to make mistakes in its identification of paradigms and in its predictions of the consequences of alternative courses of action. Such errors of scientific judgement are becoming all too frequent. Loss of trust in science by users of science is liable to result in retardation in the practical application and beneficial utilization of research findings. Many examples are now resident within pig production, including overuse of exogenous growth promoters and feed additives, excessive physical restraint of animals, introduction of fully-slatted floor and slurry-based housing systems, many forms of fully environmentally controlled buildings and so on. Indeed, the whole of the multimillion pound investment over the last 50 years into the determination of nutrient requirements has been put at risk by a failure on the part of some members of the scientific community to appreciate that the identification of nutrient requirements for the various biological functions is not synonymous with writing a rule book of recommended nutrient allowances. Bald directives from science for inclusion rates of dietary nutrients are only likely to lead many animal feeders into getting it wrong for much of the time; whereas statements of the causal relationships between nutrient input and product output allow individual feeders to calculate required nutritional strategies for their own particular environmental conditions and production circumstances.

Information, knowledge and wisdom

The base requirement for science is information, often presented in the form of descriptions of events, observations of the animal system, documentation of experience and comparisons of treatments. In animal science, it is remarkable how often these descriptions are made in terms of weight and/or time. Information is the raw material of knowledge, but is often mistaken for knowledge itself. It is also often erroneously assumed that merely by increasing the mass of the information, this will – in some mysterious way – turn information into knowledge. Knowledge stems from the synthesis of information into a cohesive view of animals and their behaviour in response to particular internal or external environmental pressures. It is often associated with an understanding of causal forces and an ability to quantify relationships between cause and effect. For information to become knowledge usually requires a hypothesis and frequently a model or metaphor to allow the phenomenon under review to be handled by the limited range of the human intellect. Many of the descriptions and explanations used in animal science are metaphors in part or in whole; for example, the nutrient content of feeds, descriptions of nutrient use (especially energy), aspects of inheritance, assessments of animal behaviour and welfare, and principles of growth and development. It is largely the ability to handle models and metaphors, to synthesize knowledge from information, to create hypotheses, to test, reject, reassess and recompile ideas, that distinguishes the scientist from the technician.

Wisdom comes from the proper use of the knowledge gained. Animal science is fortunate in relating to an industry and to the real world. This world outside science expresses elements of politics, finance, market forces and human psychology. The forwarding of knowledge within a political ethos in a way that ensures its acceptance rather than its rejection, has as much to do with scientific advance as the generation of the knowledge in the first place. Ideas will not be taken up by an industry unless there is understanding of the psychology of innovation and of the inception of new ideas. Again, market forces and finance influence the use of science by forcing decisions as to its cost-benefit. While all scientists do not need to be wise in this context, one member in any team needs to be if the results of experimentation are to be promulgated effectively and the team well supported in the future.

Research management

Research managers are charged with the task of ensuring that the experiments conducted are the experiments needed. Identifying needs for research activities requires to go much further than the identification of problems needing solutions. Simplistic approaches to problem solving tend to result in applied experiments whose design copies the nature of the problem. It would be wrong to assume, however, that the protein requirements of pigs are best investigated by offering pigs a series of diets with a range of protein concentrations. Quite the reverse, this particular issue is only open to proper elucidation by fundamental research relating to nutrient absorption and utilization. There is currently abroad the view that research management means tight control, and a close relationship between types of problems and types of experiments. There can be dangers in the idea that research management structures should be raised for animal research as if their organization was to be directed toward the production of manufactured goods; in this case, data-sets of statistically significant results.

Industrial management relies on agreement as to shared physical objectives – usually financial investment and wages on one side, and goods produced on the other. Scientific management requires a concensus, not so much of physical as of metaphysical objectives; the parties involved must agree a philosophy before a production schedule. The product, in this case scientific advance, is closely identified with the ego of the primary producers – the scientists. It is a part of them, and like their children carries forward their genes into the wide world and future generations of human endeavour. The creation of good science therefore requires a particularly sensitive and personal touch on the part of the research manager. It requires a fundamentally different management approach and management structure than, for example, the production of manufactured goods. Command lines need to be short, and teams small and preferably geographically replicated. The manager needs to be able to associate closely with the detail of the work of the managed. Such a structure has been well accepted in university circles for many years, where junior postgraduate students often share a bench with their research-project director.

Subjectivity in animal science

The pig sciences, amongst the animal sciences, are quite outstandingly active. This reflects the dominating importance of pigs in Europe, northern America and in very many other areas of the world as a provider of meat. But it also reflects the particular qualities of pigs as excellent experimental animals, and the dedication and enthusiasm of pig scientists. The total number of world pig research workers is, however, few and a remarkably high proportion are well known to each other.

Often, the lay public believe that a part of good science is the objectivity of the scientist towards the conduct and outcome of his research programme. Such objectivity is perhaps possible with unthinking pursuit of information by means of empirical trials. It is also possible if the furthering of the particular part of science in question relies solely upon receipt of additional observations of biological actualities. However, if hypotheses are to be raised, if proposals concerning causal forces are to be offered, if experiments are to be designed to test systems at their points of assumed sensitivity, if the interpretation of results is to go beyond technological number crunching, and if science is to be forwarded by proposition and counter-argument, then the assumption that scientists ply their trade with objectivism is self-evidently naive. Invariably, scientific advances are associated with the personalities and characters of individuals. The abstraction of a scientific finding from the identity of its originator is to endanger the science and to risk ossification of thought. In the animal sciences there are few facts and fewer laws, but a great deal of metaphor, concept modelling and personal opinion. That personal opinions may find themselves backed by statistical significances does not lessen the need to associate such opinions with the originator and never with the science.

There is no disgrace in the use of subjectivity in the pursuit of the animal sciences; quite the opposite, it is only by such means that real advances in knowledge and understanding will be allowed. Subjectivity in this respect should be clearly distinguished from bias, prejudice and partiality. These are the toxins of science, not to be tolerated at any cost; rooting out these evils is one of the often neglected responsibilities of research managers. The animal sciences seem to flourish best under an aegis of subjective impartiality.

Whilst at any one time the scientist must be expected to have a firm view about his science, he should always be ready to change that view and take on board new ideas with equal firmness and fervour. Scientists tend to accept quite readily that the understanding of the animal kingdom is helped if research workers take a trenchant view of their work, and present it to their peers in a clear and unequivocal way. The presenter of research findings is required to take a stance amongst his colleagues in order to allow that same stance to be effectively challenged. But unfortunately, sometimes scientists are guilty of confusing the scientific forum with the public forum. Results of research must be presented to fellow scientists even when incomplete if they are to receive early criticism and debate; but the same incompleteness may become scientific irresponsibility if research results are made freely available in a public place and to the media. All too often, scientists go public too soon. Society does not readily understand, nor warm to, the idea that scientific findings can be subsequently modified or even denied. This problem is particularly acute with matters biological. The public perception of science derives heavily from inorganic chemistry, Mendelian genetics, the laws of physics and surgical medicine. This encourages expectations of absoluteness in science; a totally inadequate expectation with which to deal with matters biological. Current public attitudes in seeking from science finite statements and clear, absolute rules for the conduct of life are apparent in society's view of issues relating to diet and health, animal welfare, and the environment. It is further abundantly clear that hard and fast rules are inappropriate ways of dealing with these issues, and attempts to formulate such rules have resulted in the scientific community now being faced with the dilemma of leaving extant what are now known to be untruths or part truths, or turning over the previously publicly accepted scientific wisdom and losing credibility thereby.

The changing nature of animal science

Over the last 50 years the objective of animal science has been twofold; first, to learn more about the animal kingdom and, secondly, to improve the efficiency of animal production. Now a third dimension has appeared and this relates to the environment within which the animals are to be used as providers for food for people, and this is closely connected to the present state of food oversupply. The components of this third dimension are broadly:

- The appreciation of consumer needs.
- Responsibilities for animal welfare
- Awareness of environmental issues.
- A disruption of the previously almost perfect relationship between the level of productivity and the level of profit.

Consumers no longer just require their food to be cheap. But price will always be an important part of the perception of value, and there is no doubt that society expects the proportion of total income that is spent on food to continue to decrease, and thus allow a greater amount of free money available for voluntary spending and investment.

Quality is now a prime consideration in the assessment of pig meat; quality is measured in terms of leanness, colour, eatability, product range and convenience in meal preparation. But to an ever increasing extent, quality is coming to relate also to the way the consumer thinks about meat. Current perceptions of healthiness go beyond the content of protein, vitamins and minerals, and extend into wholesomeness, freedom from added chemicals, and primary production in the absence of drugs, growth promoters and any other artificial aids to husbandry. Society is also becoming less sympathetic to the unacceptable consequences of intensive pig production methods; amongst these are environmental pollution and animal welfare. Pigs under the care of man have little control over their environmental welfare and have many natural choices denied them. Great thought is required in the proper provision for the housing and husbandry of pigs if their lifestyle is to be prevented from degenerating into frustration and apathy.

It is strange that it has been society's demand for cheap food over the past 50 years that created the push toward intensivism in pig and poultry production. Now, the same society finds the consequences of its action unacceptable and there is a build-up of pressure to reverse the trend. In particular, there is insistence that the welfare of pigs be improved.

Many of the aspects of modern intensive pig production are seen by society as contrary to welfare:

- Density of stocking.
- Slatted floors and lack of bedding materials.
- Barrenness of physical environment.
- Liability to diseases of intensivism, such as enteric disorders, pneumonias and viruses.
- Excessive restraint of pregnant and lactating sows in stalls and farrowing crates.

Some of the problems are more imagined than real, but it is the right of customers to make whatever demands they wish upon the contractor producing their goods. Whilst it is vital that society is educated in such a way that its choices can be more rational and informed, it is an undeniable facet of contemporary animal science that consumers are exercising their rights to choose and much of future research will be directed towards satisfying their choices.

Animal welfare issues as currently expressed seem almost more related to the needs of people than to the needs of pigs. Calls for a reduction in restraint and an increase in space result from the simplistic and erroneous assumption in the mind of society that welfare is synonymous with increased freedom of movement. Animal behaviour is a young science, still struggling to establish its ground-rules. Meanwhile, some things have been learnt from behavioural studies at Edinburgh, the purpose of which is to improve the welfare of pigs whilst maintaining productivity.

The degree of welfare is probably best assessed not in relation to human preconceptions of what animals might need, but in relation to the animal's ability to cope with the environment in which it is placed. This begs the question of the definition of 'coping'. But inability to cope is likely to be manifested by extremes of lethargy and uninterest, extremes of activity and the degradation of thinking activity into unthinking stereotypes, reductions in biological performance as measured by rate of growth and reproduction, and susceptibility to ill health. On these criteria it is evident that pig welfare in modern intensive systems can be improved, but it is equally evident that simply returning to the production systems of the past are just as likely to be counterproductive and to diminish rather than improve existing levels of pig welfare. The science of animal behaviour will come to be responsible in the future not only for ensuring welfare, but also for improving the quality of husbandry and the efficiency of production. Given the funding it deserves and needs, animal behaviour will come to take its place alongside nutrition and genetics as one of the fundamental elements in the fabric of pig science.

Chapter 2 Growth of pigs in relation to other farm animals

Introduction

Growth relates to gain in weight brought about by cell multiplication (for example, pre-natal cleavage), cell enlargement (as in post-natal growth of muscle) and the incorporation of material directly into cells (as for lipid inclusion into fatty tissue).

Development relates to changes in the shape, form and function of animals as growth progresses. There are differences between strains of farm livestock species in the way they develop. For example, the blocky Piétrain, Belgian and German Landrace-type pigs develop in a way that results in meatier hams and larger loin muscles. Fatty strains of pigs will develop in such a way as to lay down more fat and be in consequence a different shape from lean strains. The study of development, that is the change in body proportions and body shape as the animal grows, has not received nearly as much attention as that of growth. The reason for this is that, as a pig develops, the differential deposition of fatty tissues in the various parts of the body is not very evident. Most pigs lay down some two-thirds of their fat externally and about one-third internally. Subcutaneous fat tends to be laid down in a way such that, although different parts of the body have different amounts of fat, these do not develop sequentially. Whilst the shoulder is invariably fatter than the loin, a young fat pig will carry a similar balance of fat in its various depots as an older fat pig.

Perhaps another reason for development receiving scant attention in pigs is because there is little difference in the value of pig meat from the various different joints. Pigs differ from beef and lamb in that there is much less of a gradient in eating quality and acceptability between the various cuts and the various parts of the body. In general, then, differential development and changes in shape as the animal grows are not especially important in relation to the production of meat from pigs.

In contrast, the absolute rate of growth is a prime determinant of the efficiency of conversion of feed into meat, while the relative rates of growth of bone, muscle and fatty tissues are important considerations in the provision of meat for human food. As the nutrient cost of fatty tissue growth is nearly five times that of lean tissue growth, the ratio of lean to fat is second only to growth rate itself as a controller in the efficiency of feed use. The ratio of lean to bone is an important contributor to carcass value, whilst the amount of fat on the meat is a vital indicator of product quality and an important part of carcass classification, with subsequent ramifications into meat processing and retailing practices.

The relationships between bone, muscle and fat growth will, of course, also affect the development of the animal inasmuch as, at any given weight, animals grown slowly and being of greater age will tend to be leggier, have larger heads and appear rangy. Animal body shape is greatly influenced by the degree and position of fat cover. The rounded 'meaty' shape of a sucking pig, or a well finished Aberdeen-Angus bullock or Suffolk lamb, is likely to be due to a high level of covering subcutaneous fat, although this is not invariably the case, as for example in double-muscled beef animals, and the Piétrain and Belgian Landrace-type of pig, in which a well rounded rear quarter is indicative not so much of fatness, but indeed of a high lean-meat yield.

Growth occurs by the accretion of bone, fatty and lean tissues in the body. It is the result of a positive difference between the continuous anabolic and catabolic processes associated with tissue turnover. Because most fatty tissues (except brown fat) turn over slowly, there is a close association between the absolute amount of fatty tissue anabolized and the absolute amount accreted. On the other hand, lean tissue turns over rapidly, such that accretion may be only 5–20 per cent of the total protein anabolized (the proportion depending upon the degree of immaturity).

Samuel Brody, in his classical book *Bioenergetics and Growth*, published by Reinhold (New York) in 1945, says:

> The age curve of growth may be divided into two principal segments, the first of increasing slope, which may be designated as the self-accelerating phase of growth, and the second of decreasing slope, which may be designated as the self-inhibiting phase of growth. The general shape of the age curve may thus be said to be determined by two opposing forces: a growth accelerating force and a growth retarding force. The former manifests itself in the tendency of the reproducing units to

reproduce at a constant percentage rate indefinitely, when permitted to do so. In the absence of inhibiting forces, the number of new individuals produced per unit time is always proportional to the number of reproducing units. That is, the percentage growth rate tends to remain constant.

These remarks are placed in the context more of population growth and growth of cells, rather than whole-animal organisms; but Brody also makes the point that he believes that the two can be fitted to the same conceptual frame (see Fig. 2.1).

Figure 2.1 The sigmoid growth curve; weight over time

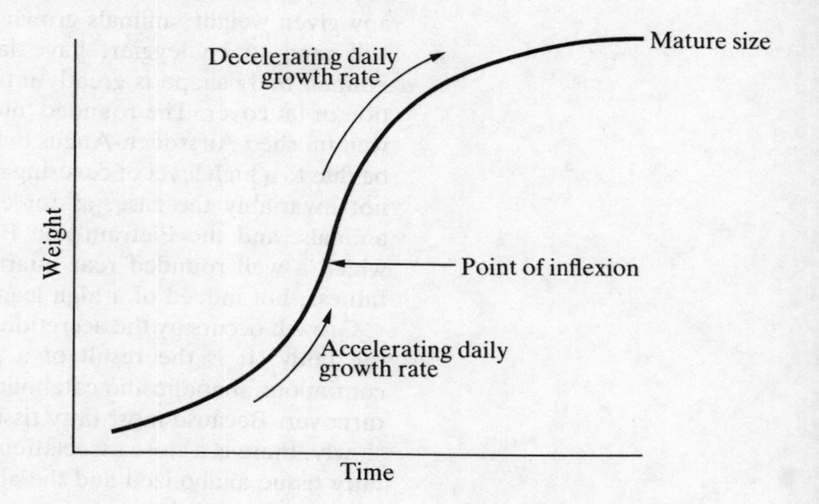

Pomeroy, 10 years later, in *Physiology of Farm Animals*,* develops the theme further,

The curve of growth produced by plotting weight against age is sigmoid . . . The general shape of the growth curve is produced by the interaction of two opposing forces, a growth accelerating force and a growth retarding force. When the slope of the growth curve is increasing, the growth accelerating force is predominating and when the slope of the growth curve is diminishing, the growth retarding force is predominating. . . .

* *Hammond, J. (Ed.) (1955) Progress in the Physiology of Farm Animals, Vol. 2. Butterworth, London.*

> The part of the growth curve during which the growth accelerating force is dominant is sometimes referred to as the 'self-accelerating phase of growth'. . . . Where the two segments of the growth curve represented by the self-accelerating and the self-retarded phases of growth intersect is a point of inflection. This point of inflection represents the point in the growth curve when the acceleration of growth has ended and the retardation of growth is about to begin, and it is therefore the point at which growth rate is at a maximum. This inflection . . . is often referred to as the pubertal inflection . . . In most of the higher animals puberty occurs after about 30 per cent of the mature weight has been achieved . . .

This view of growth has led to the proposition that maximum growth rate in farm animals cannot be achieved until the end of the self-accelerating phase of growth, around about puberty, and between about 30–50 per cent of mature size.

It now seems likely that this view is unhelpful in terms of understanding the growth process and maximizing the productivity of our farm animals. It can be agreed that embryonic and early foetal growth must be self-acceleratory on account of there having to be some relationship between current mass and accretable mass; and growth to maturity must indeed be deceleratory if mature size is to be rationally attained. But the intervening period need not necessarily be a mathematical continuation of these two phases. This intervening period is of prime importance for the production of human food as it spans the time between birth and slaughter, which is usually at 50 per cent of mature size or less for our domestic livestock species.

Although sigmoid growth curves may be readily drawn, for example for chicks and calves weaned on the day of birth, it can be proposed that the self-accelerating phase of early post-natal growth should properly be attributed to failure to supply adequate nutrients. Indeed, where nutrition and environment are unlimited there is evidence that the absolute rate of growth can be highest in the young animal and, quite conversely to self- accelerating, the actual pattern of growth may be a reduction in absolute growth rate from a peak attained soon after birth. Examples might be the early growth of the seal and of single suckled lambs. Given unlimited nutrition and environmental circumstances, it may be that the maximum absolute rate of growth may be achieved early in life and may be relatively constant through most of the growth period. Such a proposition denies the idea that achievable growth rate is somehow related in a constant way to the mass of the body already attained. Also denied is the corollary to the sigmoidal growth curve, which is that daily gain as a function of age or weight is a quadratic response, peaking at around about half of mature weight.

The proposition that the instantaneous relative growth rate (k) is a constant is consistent with the idea of self-acceleration; the greater the current weight, the greater the daily gain. Relative growth rate is a function of the growth already made. The new proposition has to be that k is not constant and the relative growth rate is not a function of the growth already made, but a function of some other attribute. As weight increases, k diminishes smoothly, in order to bring about linear rather than exponential absolute growth.

Brody's instantaneous absolute growth rate allows a look at what is happening in a very short time period, whilst his more used instantaneous relative growth rate expresses the gains per unit time (a very short period of time) as a function of current weight. When expressed as a percentage, the instantaneous relative growth rate therefore shows what percentage of the growth already made will be the next increment of growth. Following from the assumptions of sigmoidal growth, our expectations in the self-accelerating phase are therefore that a constant instantaneous relative growth rate will bring about an increase day upon day of the absolute amount of growth made. In the retardation phase of growth after the point of inflexion, then the next expected increment of growth rate will relate to an ever diminishing expectation of growth yet to be made.

Brody's propositions relate with great efficiency and exactitude to the maturing process of the later stages of life. The way animals reach mature size, in terms of their weight for age and final weight at maturity, may be clearly described (Fig. 2.2). However, it is not so conspicuously evident that the data fit Brody's assumptions in early life. There is little evidence of exponential growth and sigmoidism in the young animal. Indeed, while many descriptions of absolute weight upon absolute age show exponential growth *pre*natally, it is only with difficulty that continuation of exponential growth *post*natally can be substantiated.

Rather, with the important exceptions of the difficult periods of stress immediately after birth, when the animal is especially immature, and after abrupt weaning in the mammal, the appearance seems to be more of linearity up until the growth retardation phase is approached and the long slow haul to maturity commences. The picture of the growth of dairy cows (Fig. 2.2) shows clearly how the examination of growth to maturity may detract from analysis of early growth. Fifty per cent of mature weight is achieved within the 1st year of life; but maturity is not until the 6th year.

Figure 2.2 *Growth curves to maturity for two breeds of dairy cow and two breeds of sheep, one of each larger maturing than the other (Adapted from Brody, S. (1945)* Bioenergetics and Growth. *Reinhold, New York)*

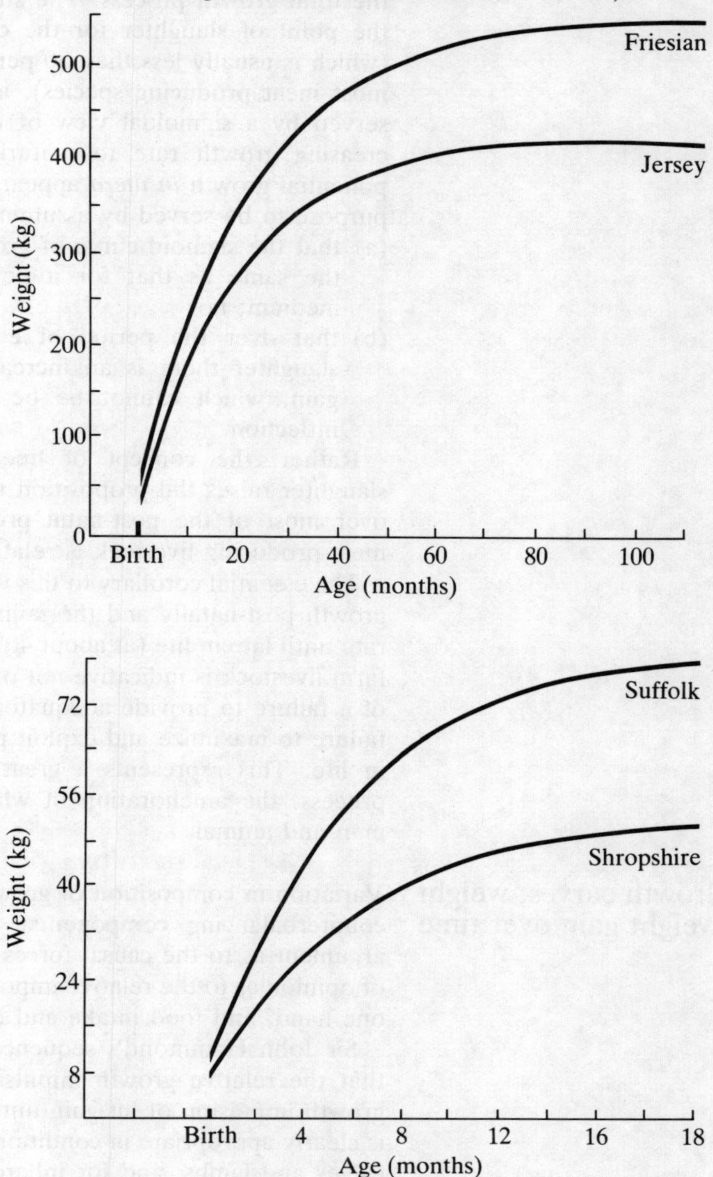

For purposes of understanding meat production from pigs, interest is confined only to the growth activities in the first part of the total growth process. The study of post-natal growth, up to the point of slaughter for the commercial production of meat (which is usually less than 50 per cent of mature body weight in most meat producing species), is therefore not necessarily best served by a sigmoidal view of the growth process. Whilst decreasing growth rate to maturity has to be correct and exponential growth *in utero* appears logical, there may be no good purpose to be served by assuming either:

(a) that the sigmoid curve of growth for an individual animal is the same as that for a population of cells on a culture medium; nor
(b) that over the period of economic growth from birth to slaughter there is an increasing potential for daily weight gain, which cannot be be maximized until the point of inflection.

Rather, the concept of linear growth between birth and slaughter raises the proposition that the absolute rate of growth over most of the post-natal productive period of life for our meat-producing livestock is relatively similar day on day.

The essential corollary to this view is that observed exponential growth post-natally and the failure to achieve maximum growth rate until late in life (at about 40% of mature size) in commercial farm livestock is indicative not of any natural law of growth, but of a failure to provide adequately for the animals' needs, and a failure to maximize and exploit potential growth to be had early in life. This represents a great inefficiency in the production process, the amelioration of which would greatly benefit both man and animal.

Growth curves; weight and weight gain over time

Variation in composition of growing animals is centred upon the counterbalancing components: bone, lean and fat. There is argument as to the causal forces of the variation, and difference of opinion as to the relative importance of time and weight on the one hand, and food intake and diet quality on the other.

Sir John Hammond's sequence of growth (Fig. 2.3) proposes that the relative growth impulsion of muscle gain exceeds the growth impulsion of fat gain until muscle slows down. This view is clearly appropriate in conditions of limited nutrition of weaned calves and lambs, and for inherently low-appetite avians, but is

Figure 2.3 Sir John Hammond's analogy to describe the sequence of growth. (From Hammond, J. (Ed.) (1955) Progress in the Physiology of Farm Animals, *Vol.2. Butterworth, London)*

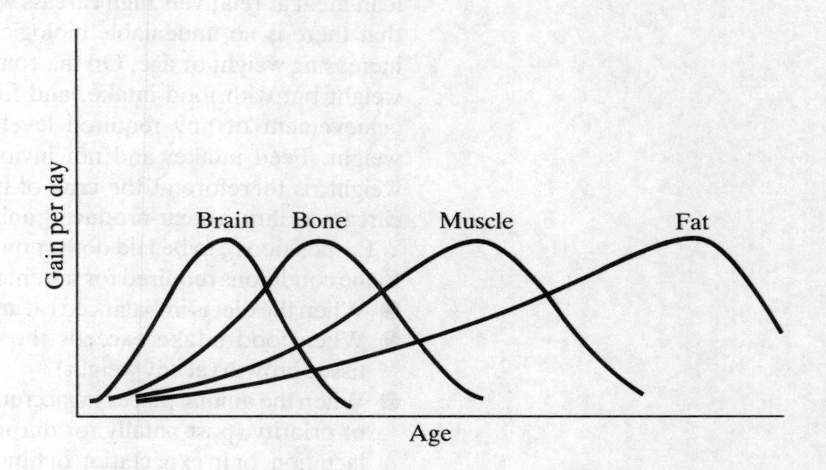

Figure 2.4 Showing how pigs become fatter with age in an exponential way. (McMeekan, C.P. (1940) Journal of Agricultural Science (Cambridge) **30***: 276)*

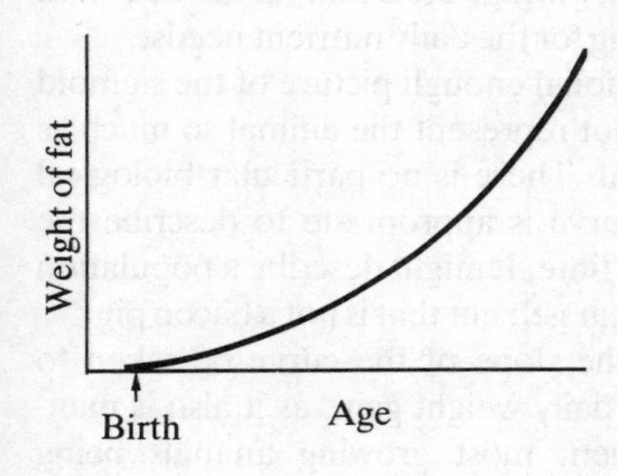

less well supported by examination of growth in circumstances of abundant and unlimited nutrient supply, such as enjoyed by very young sucking pigs. Little pigs gain fat rapidly in early life; and their most rapid fat growth impulsion may well have occurred before 4 weeks of age, whereas maximum lean impulsion may not be obtained until 12 weeks or later. Similarly, lambs fatten rapidly whilst sucking; after weaning, body composition does not change much until final fattening out in readiness for market (presuming that the market does demand fattened lamb).

Whilst the percentage composition of fat in a pig is less than 2 at birth, by 21 days of age there is usually more than 15 per cent of lipid in the baby pig's body. Modern pigs are never fatter than when they are weaned at 21–28 days of age. They have thus attained maximum fatness in the course of the first month of a 5-month growth time to slaughter and a 10-month growth time to mature size. Carcass pigs can achieve slaughter weight at 90 kg with less than 10 per cent of fat in the carcass, with the consequence that they may soon rival broiler fowl as being the leanest meat available for human consumption.

McMeekan's view (Fig. 2.4) of fat weight increasing exponentially with age from birth must therefore be disregarded as representing any age-related biological law. However, as an observed phenomena, that many of our farm animals get fatter as they grow bigger when fed fully to appetite remains indisputable and is a result of appetite increasing faster than the animals' ability to grow lean. Modern requirements for lean meat at relatively high carcass weights has nevertheless taught us that there is no undeniable biological law of increasing fatness with increasing weight or age. On the contrary, the relationship is not with weight but with food intake, and food intake control will cause the achievement of any required level of fat at any required carcass weight. Feed intake, and not inviolable rules relating to time and weight, is therefore at the crux of body composition change and of carcass quality in meat-producing animals.

Propositions can be laid down now – and returned to later – relating to the conditions required for fattening to occur. These are:

- When the diet is imbalanced (at any weight).
- When food intake exceeds the needs of maintenance and lean tissue growth (at any weight).
- When the animal places fat accretion above lean growth in its order of priority (post-natally, or during pregnancy when preparing for lactation, or in expectation of times of food shortage).
- When mature lean mass is achieved and ingested food has no other function than fat growth to satisfy.

There is every reason to surmise that McMeekan's (1940) pigs had high appetites and low lean growth rates: a greed-fostering combination not found in modern slim-pigs of low appetite and high lean growth; nor for that matter to be found in the parsimonious broiler fowl, or ruminants given low-quality forage diets that fill the belly and satisfy the appetite before providing for the daily nutrient needs.

Figure 2.1 represents a conventional enough picture of the sigmoid growth curve, but one that may not represent the animal so much as the hand of man upon the animal. There is no particular biological reason for believing a sigmoid curve is appropriate to describe the increase in mass of an animal over time. It might describe a population of bacteria growing on a plate of agar gel; but that is not a bacon pig.

Neither may the gradient of the slope of the curve be taken to represent any biological limit for daily weight gain, as it also is man-created rather than animal-driven; most growing animals being nutritionally restrained by the actions of their keepers. Nutritional restraints seem to be least in poultry and most in ruminants, with pigs intermediate. Only when fed as if they were monogastrics might ruminants move into the region of nutritionally unrestrained growth;

examples would be intensive milk or cereal-fed beef and lambs.

Weight over time curves – if they are used at all – are probably best restricted to the lean tissue mass rather than to total live weight. Fatty tissue is variable – being so open to environmental and nutritional manipulation – and some would argue almost independent of physiological growth, having a function more akin to that of a larder. Fatty and lean strains of rats given diets of widely differing energy:protein ratios all grow protein at a similar daily rate – all maximize their potential daily lean tissue growth rates; targets for lean growth dominating appetite. Body fat levels vary hugely and reflect the drive to lean growth regardless of the diet make-up or the consequences for the fatty body tissues (about which the animals seem to care little).

Figure 2.5 Showing the place of birth in relation to 'sigmoid' growth

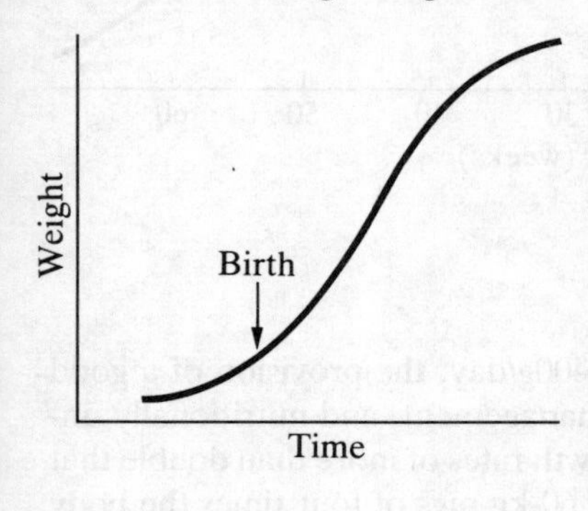

The bottom of the growth curve represents lost possibilities and it is worth reassessing whether or not it really is unavoidable and what intrinsic driving forces are at work to create observed exponential increases in growth up to the linear phase. The 'self-acceleration' rule is that the potential for absolute gain is a function of increasing tissue mass. This rule states that it is acceptable to consider a 50-kg animal growing 1 kg of lean tissue per day, but it is not acceptable to consider a 5-kg animal doing the same. Whilst at the extremes this logic must clearly hold, observed phenomena are usually not within range of it. There is a fundamental anomaly in pig farmers being disappointed if their 80-kg pigs do not grow at more than 1 kg/day, while beef farmers are delighted if 400-kg steers do as well. There are unlikely to be genuine species differences for the rules that govern the proportion of the total body lean mass that can be added to each day with new lean growth. Different rules are not evident for the energy costs of maintenance, fat deposition, protein deposition and cold thermogenesis. Many of the fundamentals of growth are common across species.

Rather, it seems as if man's school report for classes in animal care is 'could do better'. Weaned livestock get under way slowly and the curvilinear phase for youthful growth is quite evident. However, recent experiments have caused a review of this concept of early animal growth as reflecting an inherent rule. With good husbandry, linear growth can be brought back in time and encroach deeply into the assumed curvilinear phase.

If pigs are given enough to eat, it is quite difficult to prove that they grow more slowly when young than when older. For example, while observations of pigs on farms would lead one to believe that at 15-kg a

Figure 2.6 Showing the skewed nature of the original data describing weight change per unit time. (From Hammond, J. (Ed.) (1955) Progress in the Physiology of Farm Animals, *Vol.2. Butterworth, London)*

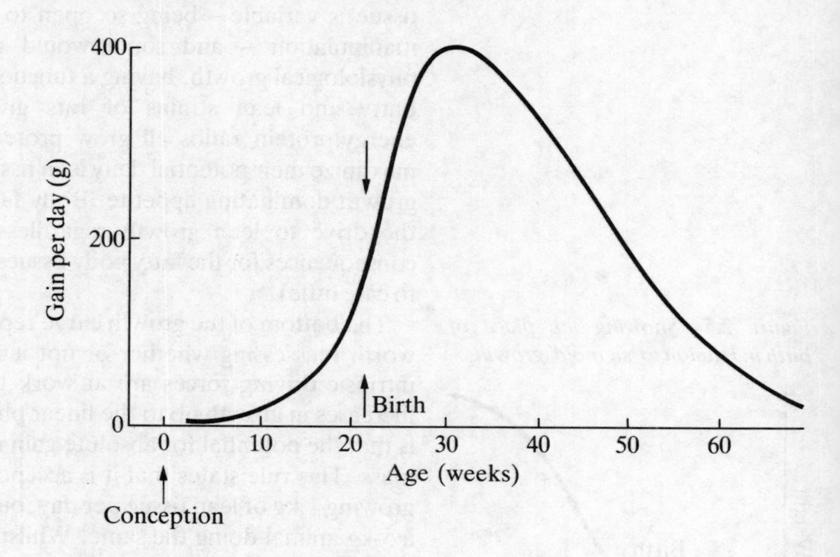

little pig could only grow at most 400g/day, the provision of a good environment, perfect health and management, and nutritionally unlimited circumstances will elicit growth rates of more than double that and similar to the achievements of 60-kg pigs of four times the body weight.

The curve is open to misinterpretation if it is forgotten that the exponential growth phase relates mostly to embryonic and foetal growth (Fig. 2.5). The confusion is perhaps forgivable in that the same sigmoid curve *may* indeed be observed post-natally, but this is consequent upon an inability to look after properly young mammals under man's care and not a consequence of any biological time-based constant.

The curves of the classical workers on growth theory, Wallace, Pálsson and Vérges, are highly skewed (Fig. 2.6), quite purposefully showing the most rapid increase in gains to occur immediately after birth. For these suckled lambs, pre-natal and early post-natal growth shows rapid and linear increases in gain to peak growth of 400 g as early as 56 days after birth (30 weeks after conception). The rapid early descent from peak in such immature animals is most surely not

Figure 2.7 *Growth of single- and double-suckled lambs (all born as singles). (Interpolated from Hammond, J. (1940)* Farm Animals. *Edward Arnold, London)*

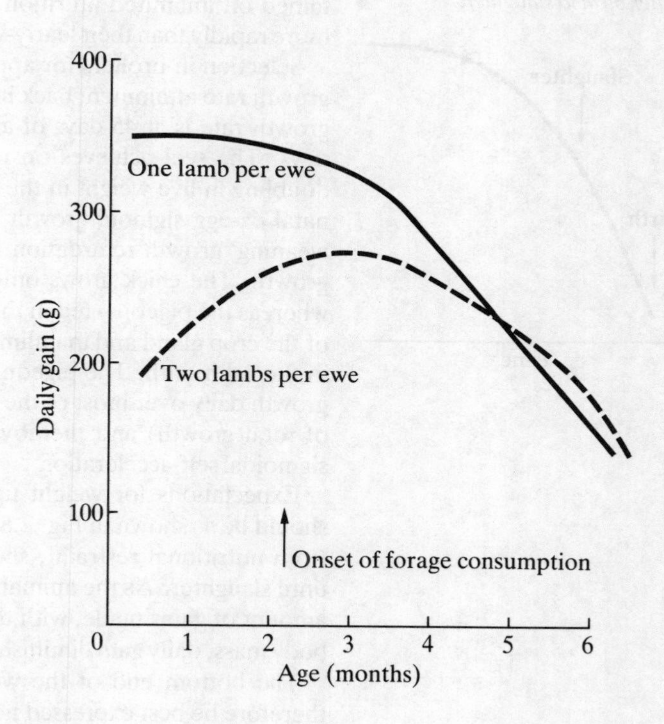

the approach of maturity; more likely it is the onset of increasing nutritional deprivation as the monogastric milk-fed lamb turns into a ruminant forage-fed young sheep.

As also witness Fig. 2.7, contrasting times for peak growth in single- and double-suckled lambs (all born as singles), and the consequences of the onset of natural weaning. Single-suckled lambs never grow faster than in the first 2 months of life, while the limited milk supply to double-suckled lambs pushes peak growth rate back to the 3rd month after birth. The human baby also conforms to the same pattern, growing most rapidly of all from birth to 6 months of age. Between the

Figure 2.8 Potential relationship between weight and time, showing the rate of gain per day to be similar from soon after birth to slaughter.

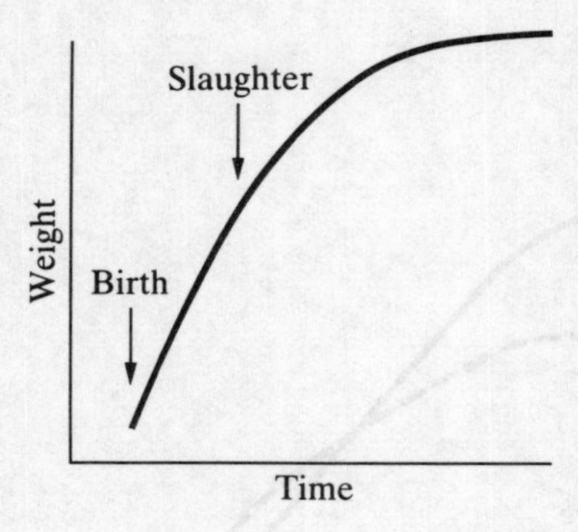

4th and 7th week of age one baby (RJW), when of 4 kg average weight, gained 290 g weekly; a rate of growth that if maintained would result in a 24- stone (154 kg) 10-year-old. Milk-fed veal calves maintained on unlimited nutrition grow very like suckled pigs and much more rapidly than their 'early-weaned' contemporaries.

Selection in broilers for appetite and growth rate has moved peak growth rate attainment back in time. Thus for 1972 strains the peak of growth rate is at 45 days of age, while for 1984 strains peak is at 35 days. The seal achieves on mother's high-fat milk an unparalleled doubling in live weight in the first 3 days of life. Chicks show a post-natal ex-egg sigmoid growth curve with all the signs of early 'post-weaning' growth retardation followed by more gently sloping linear growth. The chick grows only 150 per cent in the 1st month of life whereas the pigeon – fed in the nest by its parents from the secretions of the crop gland and in unlimited nutrition – grows over 300 per cent in the 1st month. The pigeon achieves the same absolute amount of growth daily over most of the growth period (between 10% and 75% of total growth) and thereby fails comprehensively to demonstrate sigmoidal self-acceleration.

Expectations for weight upon time between birth and slaughter should be as shown in Fig. 2.8, and potential daily rates of gain, freed from nutritional restraint, should be similar from quite early in life until slaughter. As the animal grows there is similarity in the absolute amount of gains made, with the corollary that, as a proportion of the body mass, daily gain diminishes steadily from early in life.

The bottom end of the weight upon time curve (Fig. 2.9) may therefore be best expressed not as an exponential increase in classical sigmoidal mode but as a straight line; the difference between the two representing the consequences of nutritional inadequacy and a measure of lost growth potential. This, of course, is quite at odds with the view of the 'sigmoidal school', that the ability of animals to deposit protein is a quadratic function of live weight peaking at around about 50 per cent of mature size. Whilst it may be obvious that a 1-kg animal is unlikely to grow 1 kg/day, it is *not* self-evident that a 10-kg animal cannot grow 1 kg/day and it *is* self-evident that the same animal at 50 kg will not be growing at 5 kg/day.

Conventional wisdom takes it as axiomatic that response line 1 in Fig. 2.10 represents expected maxima for lean tissue growth rates as live weight increases. But under conditions of full feeding, the possibilities for lean growth look far more like response number 2. If this is the case, then young animals following line 1, as they so often do, are

Figure 2.9 *Possible growth curves in early life: conventional wisdom favouring b, most production units achieving a, while the potential available is described by c.*

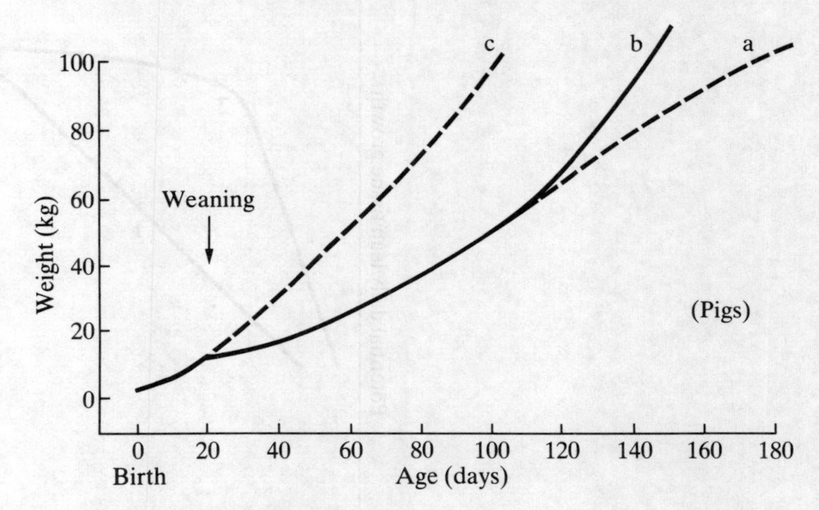

undoubtedly growing considerably more slowly in early life than they readily could. This gives rise to the proposition that 'observed rates of growth are not indicative of biological performance potentials, but rather the presence of adverse nutritional and environmental conditions'. Most young animals, especially pigs, sheep and cattle, would grow both faster and differently if they had their own way in the matter.

Analysis of weight against time (Fig. 2.11) has many insights for the production of meat from farm animals. With time, a population of animals conforming to case 1 will, in unlimited nutritional and environmental circumstances, maximize genetic potential and proceed by means of linear growth to target mature lean mass characteristic of species and breed type.

Enhancement of growth rate is a basic goal in animal production; to do so not only improves the rate of throughput and therefore the turnover of the business, but it also brings about improved biological efficiency – more meat for less food. By genetic selection, or by breed substitution, populations of animals may be obtained with faster potential rates of growth; that is to say, with steeper slopes for their

Figure 2.10 Potential lean growth rate as a function of animal age, weight or degree of maturity

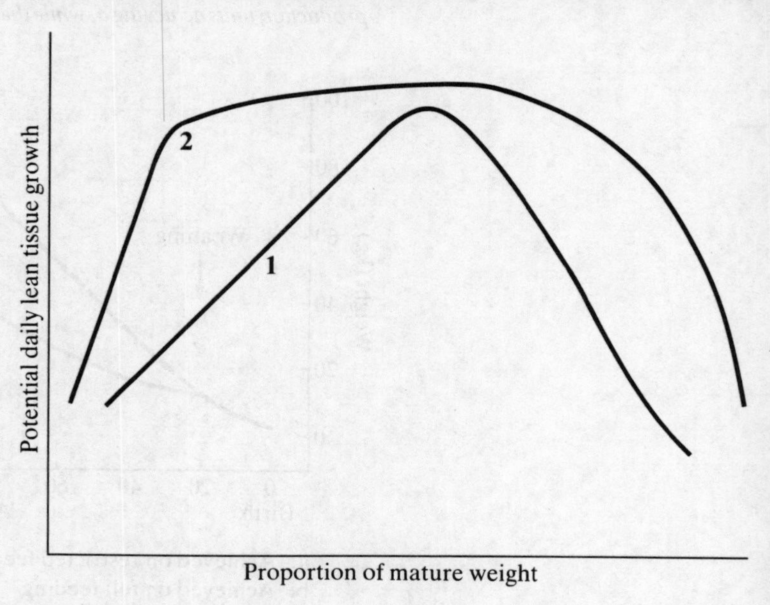

Figure 2.11 Relationships between time and mass at maturity and the consequences for growth rate.

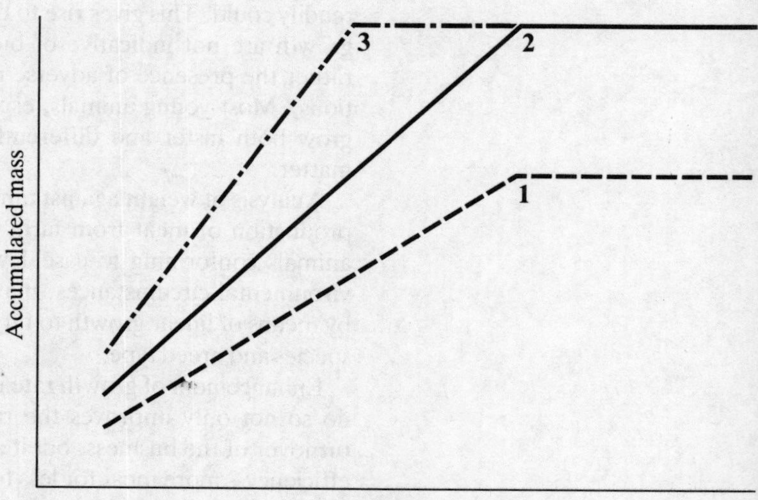

linear response of weight upon time. Case 2 shows such an animal type. If, as seems reasonable, time at maturity does not change much, then the undeniable consequence is that faster growth rate gives greater mature size. Although evidence can be brought forward to point to small differences between breeds of animals within a species in their time of maturity, the general rule must be that age at maturity does not alter much and not nearly as much as size at maturity (consider Aberdeen-Angus and Charolais cattle, broiler breeders and laying stocks, Large and Middle White pigs, Wolf-hounds and Dachshunds). The corollary is clear: by selection and breed substitution, the agricultural industry has increased the rate of growth of its animal stocks and in so doing has increased animal size. Typical examples would be the replacement of the Aberdeen-Angus by the Charolais as our major beef breed, the increase in size of Large White pigs over recent years and the abandonment of the Middle White. (As an aside, the Aberdeen-Angus is known in agricultural scientific parlance as early-maturing and the Charolais as late-maturing. By early-maturing what is actually meant is small-maturing and by late-maturing what is meant is large-maturing. The Hereford (known as early-maturing) and the Charolais (known as late-maturing) are, in fact, equally maturing in either real time or proportion of mature size. The Aberdeen Angus is, in fact, small-fattening and, if anything, late-maturing in time. The Friesian is large-fattening.)

Populations of animals conforming to case 2 in Fig. 2.11, as well as growing faster and maturing bigger, have higher maintenance requirements in their adult breeding herds and are more difficult to fatten. It has been considered, by everybody until recently and by many even now, necessary to fatten beef and lamb before they are sold for meat. Small-maturing animals can be fattened after mature size is attained and still have carcasses of handleable size – such cannot pertain for large-maturing types. Under modern production systems, for all breed types, fattening whilst growing is invariably the requirement. Large breeds with greater lean tissue growth rates are more difficult to fatten and the carcass will be less mature, being up to required weight at both an earlier age and a lower proportion of the mature mass. At equal fatness, an Aberdeen-Angus cross will weigh about 400 kg as compared with 500 kg for the Charolais cross. Bigger-maturing breeds of sheep are considerably leaner at 30 kg than smaller-maturing breeds and the same consequence follows from breeding faster-growing strains of pigs. In a market environment demanding lean meat, one may expect these characteristics to be regarded as beneficial. However, some wholesalers of some meats – not consumers – require fat on the carcass, as it facilitates handling and cutting: for example, beef and especially sheep meat. It is also true to say that some level of fat is needed for eating quality and acceptability, and in some strains of pigs over-leanness is now an increasing problem.

The animal growing more rapidly will need more food to fatten whilst growing. Thus while case 1 animals (say, Aberdeen-Angus or an unimproved pig strain) are growing relatively slowly and may therefore be readily fattened, within their appetite restraint, on food of relatively poor quality, animals of case 2 (like the Charolais, or an improved pig strain, or perhaps an entire male rather than a castrate), growing more rapidly, cannot fatten unless either appetite is greatly enhanced or a specially high-quality food is given. With regard to pigs, this has created a need to increase the intake and the nutrient specification of pig diets, while for beef it has resulted in a need to substitute forages with cereal concentrates in many beef production systems, with the rather strange consequence that, in order to fatten modern strains of beef animal in intensive systems, supposedly forage-using beef cattle may actually use more grain per kilogram of meat produced than pigs or poultry. Although fatness used to be an asset in meat, it is now unwanted by today's consumer. There should therefore be no need to *fatten* livestock at all. However, there is a need for further development work to allow intermediate trades and meat packers to take on board the new technologies needed to deal with very lean carcasses.

Case 3 is the jewel in the crown. The animals grow fast but, being early-maturing, also mature smaller at a lighter weight. This population carries twin attributes to the benefit of both the slaughter generation and the parent breeding population. Unfortunately, case 3 is rather rare, but nevertheless should be a priority area for research endeavour and animal breeding programmes in all our meat-producing livestock.

Growth response to feed input

Whilst animal growth usually relates to a base of time, weight change is a direct response to feeding.

Feed intake – and not inviolable rules relating to weight, time and sigmoidism – is the crux of body composition change and of carcass quality in meat-producing animals. The issue is perhaps best considered in terms of tissue weight per unit time: the daily growth responses of lean and fatty tissues to daily inputs of food. If growth is to be understood, then there must be quantitative descriptions of these

Figure 2.12 Showing the linear response of protein retention to feed intake, with no curvature even in these very young pigs. (Note the high feed intake achieved.) (Calculated from Campbell, R. G. and Dunkin, A. C. (1983) Animal Production **36**: *185)*

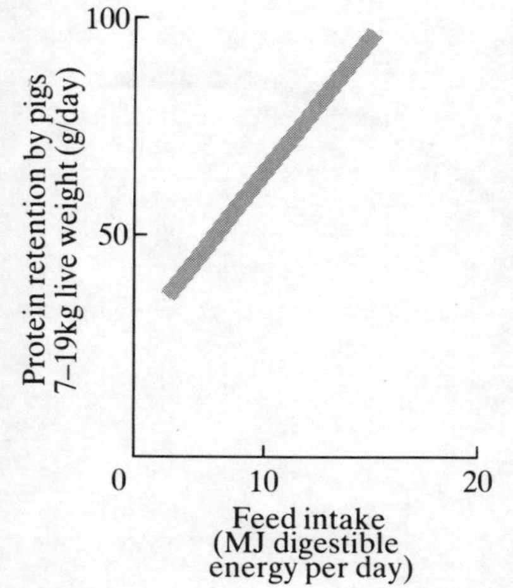

relationships and hypotheses as to their controlling forces. There is general agreement on such matters as the energy requirements for maintenance, for the energy cost of the deposition of protein, energy for the deposition of fatty tissues, for cold thermogenesis and so on. There is also general agreement about protein needs for maintenance and new tissue synthesis. There is some level of consensus about protein turnover rates, and relationships between total protein synthesis and new protein accretion. But there is controversy about the controlling forces of animal growth responses to nutrients and consequently a need for some hypotheses.

In young pigs the response of protein retention to feed intake is linear up to the maximum of appetite (Fig. 2.12); while in slightly older animals (Fig. 2.13), at high levels of food intake protein retention is seen to plateau. In these particular studies from Australia, unusually high intakes were achieved. Appetites of 36 MJ of digestible energy per day are more usually associated with 70 kg pigs than with those of 35 kg. It is evident from this work, which accords with studies previously undertaken in Edinburgh, that protein growth responds linearly to feed intake up to a maximum point, at which it plateaus.

Given high feed intakes, this plateau may be attained relatively early in life. And it seems (Fig. 2.14) as if maximum limits to daily protein retention are rather constant over a wide range of live weights, covering the range of 10 to 60 per cent of mature weight (or about 20 to 120 kg in the case of pigs).

Much as conventionalists may wish to believe that the potential for daily lean growth increases as animals increase in live weight, reaching a maximum at about 50 per cent of mature size, it now appears that potential daily lean tissue growth is flat over a wide part of the age/weight span and may be attained early in life. This means that a general theory relating daily gains of fat and lean to feed intake may pertain over a wide range of animal weight.

Figure 2.15, relating to male turkeys with a mature size of about 20 kg, ably demonstrates the point. Growth rate accelerates rapidly to a plateau early in life, at about 7 weeks or about 15 per cent of mature weight (not at about 40–50% of mature weight). This rate does not then further increase, but rather holds at a relatively constant 120 g or so daily through to 70 per cent of mature weight. So from 3–14 kg live weight, daily gain is a decreasing proportion of weight.

As feed supply increases then a linear response in lean growth is proposed (Fig. 2.16, which relates to pigs). A plateau occurs at the maximum growth potential for the animal, which in case A is 400 g of

Figure 2.13 *Showing the linear response of protein retention to feed intake followed by a plateau at about 35 MJ digestible energy and 125 g protein retention. (Note the high feed intake achieved.) (Calculated from Campbell, R. G., Taverner, M. R. and Curic, D. M. (1983)* Animal Production **36***: 193)*

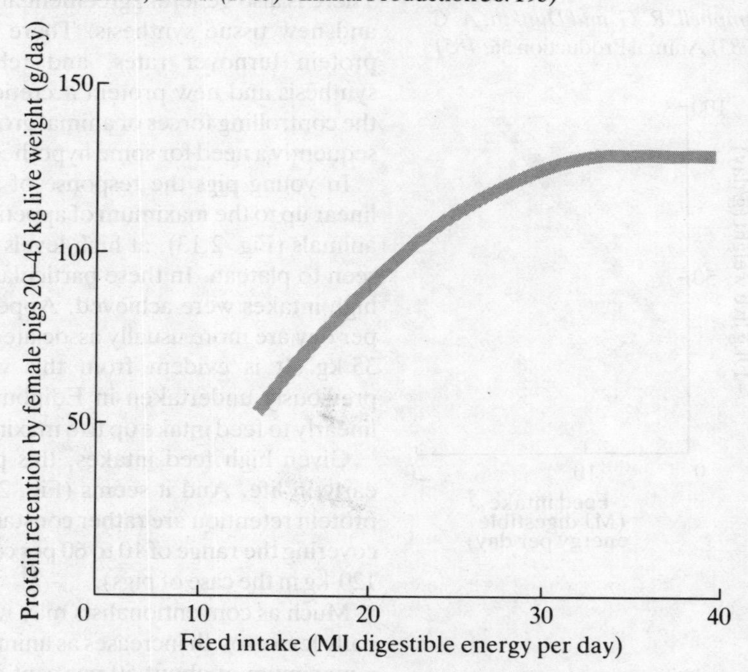

Table 2.1 Lean tissue growth potentials of pigs (g/day, total lean in whole live body)

	Potential lean tissue growth (g/day)
Entire male	600
Female	500
Castrated male	400

lean daily. In case B the maximum potential is 600 g daily. A and B may be different genotypes, or perhaps different sexes.

In pigs, the entire male has a much higher potential for lean tissue growth than either the female or, particularly, the castrate (Table 2.1). Castration results in a considerable level of potential production efficiency being lost on account of out-moded convention, and the need for meat wholesalers and distributors to adapt in order to handle the lean meat product that comes from entire male livestock. Growth rate potential in poultry and ruminants is also strongly influenced by sex; and there is a marked plane of nutrition effect – sex differences being much more noticeable at high feed intakes (Fig. 2.17).

During the linear response phase for lean growth, fatty tissue growth will be restrained to a minimum level on the assumption that, under conditions of normal growth, the animal prefers to target for lean whilst maintaining some, minimal, physiologically

Figure 2.14 *Daily protein deposition rate in relation to* mean *live weight. The experiment began at 20 kg and progressed by serial slaughter at increasing live weight points through to about 180 kg. Mean live weight is the mean between 20 kg and the serial slaughter point. (From Tullis, J. B. (1982)* Protein growth in pigs. *Ph.D. Thesis, University of Edinburgh)*

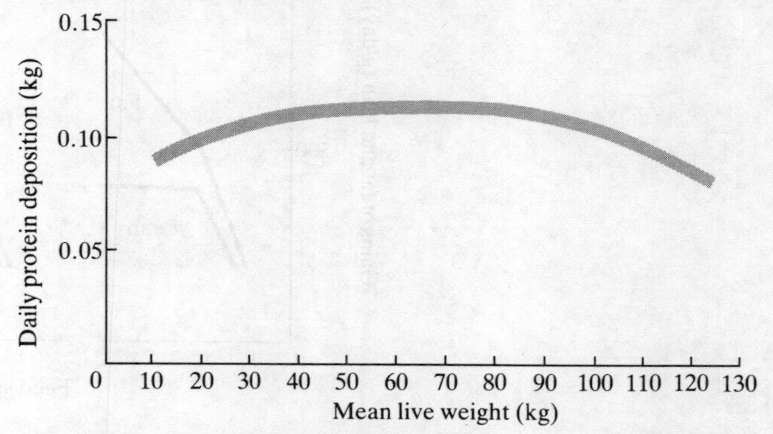

Figure 2.15 *Growth rate of turkeys as a function of live weight. The mature weight of these animals is in the region of 20 kg. (Unpublished results from Emmans, G. C., Edinburgh School of Agriculture)*

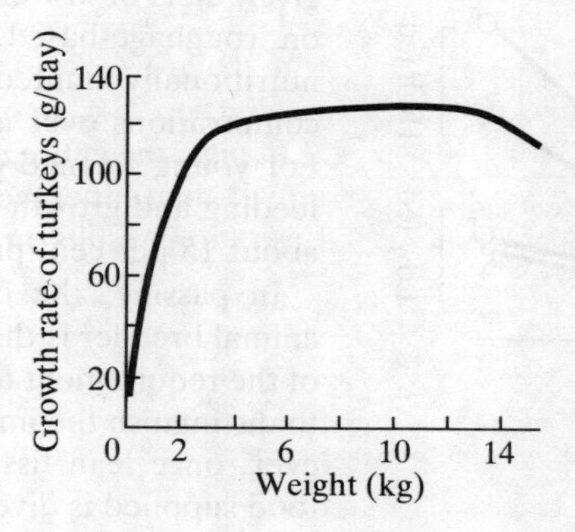

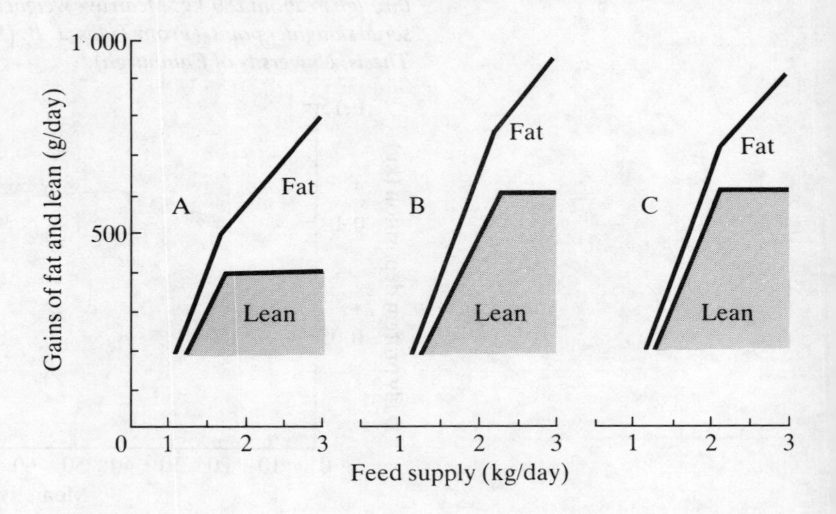

Figure 2.16 Effect of increasing feed level upon the growth of lean and fatty tissues in the whole body of pigs. (For explanation, see text pp 27–8 and below)

Figure 2.17 Effect of sex and feed level upon the growth of lambs. (After Pálsson, H. and Vergés, J. B. (1952) Journal of Agricultural Science *(Cambridge)* **42***: 1)*

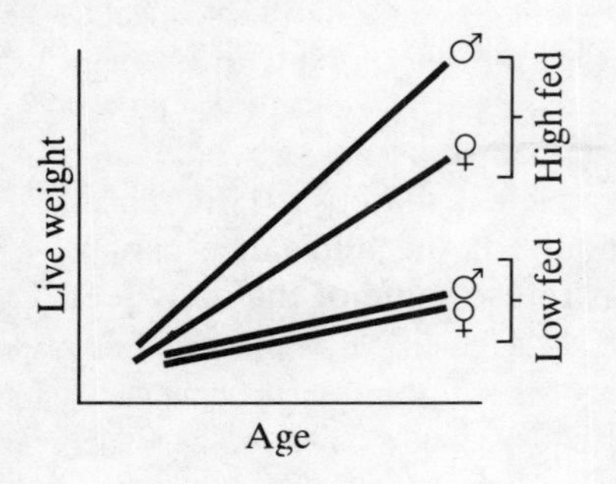

normal, level of fat in the gains (Fig. 2.16). In this case, the minimum ratio of fat to lean is about 1 of fat to 4 of lean. Until feed intake is sufficient to maximize lean tissue, the animal will not fatten whilst it is growing and there will be relative constancy in body composition over a wide range of body weight. Animals given diets of low energy density, as for example many ruminants on roughage-based diets, may be expected to remain in the nutritionally limited phase of growth and have constant body compositions over a much greater period of their life than pigs. For young weaned growing sheep there is little effect of level of feeding and growth rate upon body composition, which is usually about 15 per cent protein and 19 per cent fat.

In passing, the consequence of this view of growth for the animal breeder is that a test regime failing to supply feed in excess of the requirement to achieve maximum lean tissue growth will fail to distinguish the improved from the unimproved genotype. However, once lean tissue growth potential is attained, all the extra food supplied is diverted to fat, the animal fattens rapidly, overall growth rate decreases and feed conversion efficiency worsens. In case A, 2 kg of food per day brings about fattening, whereas for B maximum lean tissue growth has yet to be achieved. To the left of the breakpoint may be described as nutritionally limited growth, to the right may be described as nutritionally unlimited. Two kg of feed represents unlimited nutrition for A but limited nutrition for B.

It should come as no surprise to animal breeders that selection against fat so often brings about a reduction in appetite, *ad libitum* intake being pushed to the left along the feed supply scale in Fig. 2.16 (i.e. reduced) and thus becoming related to nutritionally limited (low fat) rather than nutritionally unlimited (high fat) growth. Pigs tested on an *ad libitum* feeding regime, and selected for live weight gain, would be expected to make substantial positive improvements in daily feed intake and daily lean tissue growth rate; but the pigs would also become fatter. Selection against fat, in addition to diminishing feed intake, would also decrease live weight gains and only encourage mediocre improvement in daily lean tissue growth rate. Selection for intake would result in similar responses to selection for live-weight gain, although fatness may well increase faster than lean growth. In order to balance these pressures, a weighted index can be constructed to allow simultaneous improvements in daily lean gain and reductions in fatness. Almost invariably such indices have also included feed efficiency and have resulted in by far the most pressure being against fat rather than for gains. Recently, success has been achieved using a restricted feeding scale for pigs on test. A fixed amount of feed is offered for a fixed time. The pigs begin test at the same weight (or perhaps age) and are given a daily allowance, which is increased weekly, the same amount of feed thus being offered to all pigs on test. The test finishes at a pre-set time (often after 12 weeks), and the pigs selected on the basis of weight and fatness (from which lean can also be calculated). If increased positive pressure on appetite is required, the allowance can be increased such that some of the pigs refuse some of the daily feed ration. The refusals do not need to be measured, simply cleaned away each day; animals with low appetites stand a better chance of failing to be selected on grounds of slower growth compared with contemporaries eating all the offered feed.

Strain and sex differences may also be revealed in the ratio of fat to lean in nutritionally limited growth (Fig. 2.16). In pigs, the ratio can vary from less than 0.5 of fat to 4 of lean for entire males of improved strains, to more than 2 of fat to 4 of lean for castrates of unimproved strains. Selection against fat in meat is more likely to diminish this minimum fat ratio – as the example in case C – than to increase lean tissue growth rate. Fat reduction can therefore be achieved either by selecting for animals of type C, or ensuring that feed intake never gets to the unlimited phase. Maintaining limited nutrition and thereby preventing fatness can be achieved by selecting animals of low appetite in relation to needs for maintenance and lean tissue growth, or by giving diets of low nutrient density – forages – so that the belly is filled before nutrient requirement is satisfied, or by restricting the amount of food given. A further strategy is to select directly for a low minimum fat ratio; all of which provides a further conundrum for the animal breeder trying to derive an appropriate test regime to enhance lean growth rate, improve efficiency and reduce carcass fatness simultaneously – notwithstanding the importance of maintaining meat eating quality.

The propositions put forward in Fig. 2.16 are fundamental to animal production strategies, and allow a theory of growth and fattening to embrace species. They state that there is a maximum lean tissue growth rate, and that this is largely independent of animal weight and age. They further assume linear responses up to the plateau. They govern the circumstances in which animals cannot become fat, and give rules for feeding to maximize efficiency of feed use and growth rate, whilst minimizing fatness. The relationships between appetite, lean growth potential and diet quality control fattening in a consistent way across species. Animals with higher lean tissue growth potentials can consume greater amounts of food with consequentially improved feed efficiency and no increase in fatness. Broilers do not fatten because appetite is low in relation to potential growth rate. Selecting for greater appetite will initially enhance growth and not fatness, but as lean potential is caught up with, further selection for appetite will bring about fat broilers. Pigs with reduced appetites will be lean, but slower growing, while cattle given forages will not fatten until diet quality is improved. Manipulation of diet quality and feed intake can turn pigs into fowl and cattle into pigs.

Under normal circumstances young animals will, because of limited size, have low appetites and will find themselves in nutritionally limited growth (Fig. 2.18). The same would apply for older animals that have poor appetites or that are given low-quality feeds, such that within appetite their achieved nutrient intake is low. Older animals, with higher inherent appetites, have a better chance of finding themselves in nutritionally unlimited growth. This also goes for animals given high nutrient density diets, where the proportion of forage in the diet does not limit intake.

Figure 2.18 *As animals get older and bigger, their ability to eat more food increases. This means that older pigs are more likely to become fat than younger pigs. However, if diet nutrient density also diminishes with age, then the likelihood of fattening decreases.*

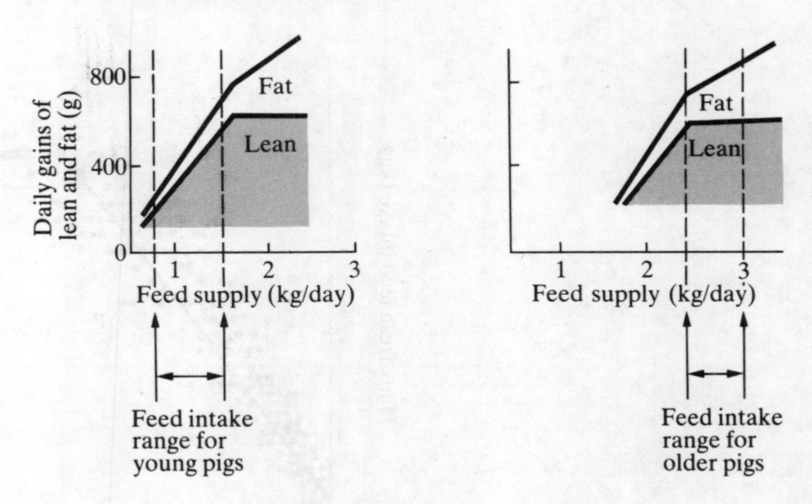

The powerful influence that feed intake has upon growth is largely due to growth phenomena being so often observed in nutritionally limited circumstances. It is quite difficult to identify times in the life of a farm animal where the achieved feed intake is, in fact, maximized. The exception to this general rule might be the young milk-fed suckler at the mammary gland, or on veal-type *ad libitum* milk-based diets.

Even with pigs it is simple to demonstrate that, under ideal conditions, feed intakes such as shown here can be as much as twice that achieved under normal farm circumstances in early life. The measurements in Fig. 2.19 are contrary to the thought that feed intake ought to be some function of metabolic body weight: not at all, it appears to be quite linear up to a variable plateau.

In Table 2.2 the feed intakes in commerce and under good conditions in Edinburgh trials may be compared with those

Figure 2.19 *Daily feed intake of pigs fed a cereal based diet of 88% dry matter to appetite. (Tullis, J. B. (1982)* Protein growth in pigs. *Ph.D. Thesis, University of Edinburgh)*
Appropriate regression equations for boars, gilts and castrated males of between 5 and 85 kg live weight were respectively:
Daily feed intake (kg)=0.046 (s.e. 0.0012) live weight + 0.35 (s.e. 0.045)
Daily feed intake (kg)=0.043 (s.e. 0.0012) live weight + 0.41 (s.e. 0.044)
Daily feed intake (kg)=0.048 (s.e. 0.0013) live weight + 0.42 (s.e. 0.053)

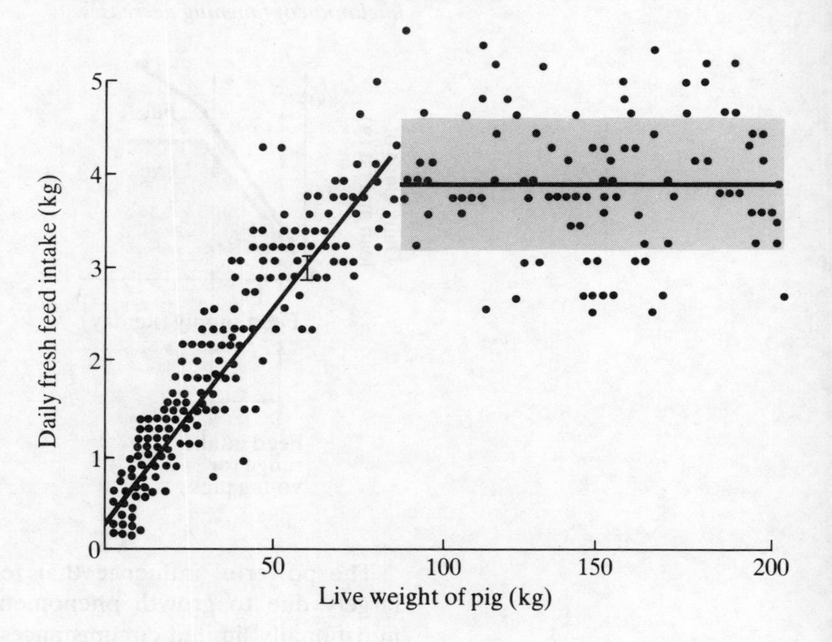

achieved by Tullis, in whose work feed intakes were practically double what might have been expected, and above the Agricultural Research Council estimates; viewed previously with scepticism, even when used for pigs of much higher live weight.

Recently, at Edinburgh, feed intakes and daily gains 50–100 per cent in excess of previous expectation or commercial achievement have been regularly achieved; for example, it is not unusual for baby pigs grown from 6 to 24 kg to eat 25 kg of feed in 31 days and grow at nearly 600 g/ day; that is, to increase their body size by 30 g/day for each kilogram of body weight.

Figure 2.20 The 'long-haul' post-weaning before attainment of linear growth (also compare Fig. 2.9)

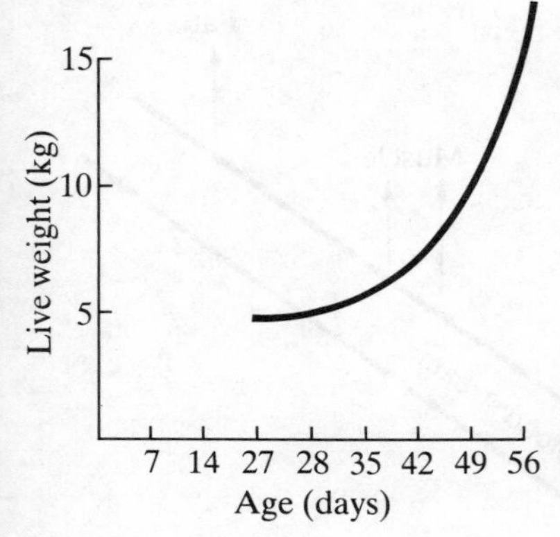

Table 2.2 Estimated and achieved feed intakes for pigs (kg fresh feed)

	Live weight of pig	
	10 kg	20 kg
ARC (1981)*	0.79	1.43
Commercial	0.40	0.80
Trials	0.60	1.00
Tullis (1982)†	0.86	1.32

* ARC (1981) *Nutrient Requirements of Pigs.* CAB, Slough
† Tullis, J. B. (1982) Protein growth in pigs. *Ph.D. Thesis, University of Edinburgh*

Failure to supply adequately the nutritional needs of the young animal after weaning can explain the curvature at the bottom end of the sigmoid growth curve and the subsequent long haul up the curve (Fig. 2.20).

Weight loss or negative growth

A series of Edinburgh experiments was initiated to examine slow growth in young animals, and also negative growth in both the young and lactating adult. Negative growth has an important contribution to make to normal growth and dramatically influences body composition.

Hammond's view of nutrient partition (Fig. 2.21) was not about positive growth – as it is so often put – but about negative growth. The proposal of 1944 was not that fat growth could not proceed until muscle growth is maximized – although that was also implied in the 'waves of growth' proposition – but that as nutrients become progressively limiting then arrows are *withdrawn* one at a time. Negative growth is not the converse of positive growth; fat anabolism may cease entirely before protein anabolism slows down, while fat catabolism and protein anabolism can proceed simultaneously.

Negative growth is a perfectly usual phenomenon, providing that it relates to fatty tissues. Thus, according to the level of feed intake in lactation, from 4 to 18 kg of lipid can be lost by high- and low-fed 150-kg sows during a 28-day suckling period. Fat loss is related to weight loss, but with a high constant, indicating that, even at sow body weight stasis (and sows usually lose about 10 kg weight in lactation), there is some 7.5 kg of fatty tissue catabolized during lactation, or 250 g/day. Similarly, in dairy cows, recently derived equations (Table 2.3) indicate lipid change to be the main contributor to change in body weight and energy content. Fat

Figure 2.21 *Hammond's analogy for nutrient partitioning. (To be found in Hammond, J. (Ed.) 1955* Progress in the Physiology of Farm Animals, *Vol. 2. Butterworth, London)*

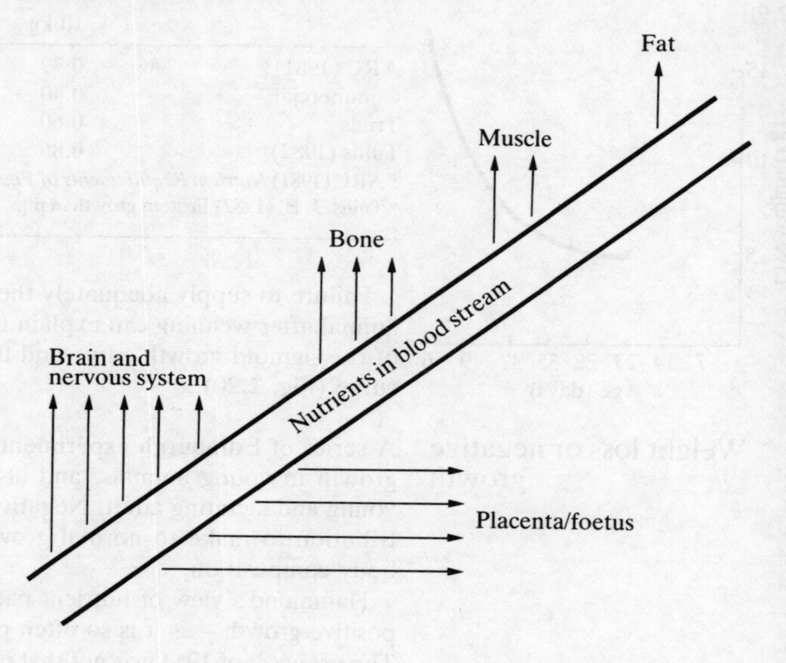

Table 2.3 *Equations from mature Friesian dairy cows*

Empty body weight $\simeq 1.1$ lipid $+ 400$
Total gross energy (GJ) $\simeq 0.041$ lipid $+ 2.0$
Lipid change is practically the sole contributor to change in body weight and energy content

losses by dairy cows in early lactation are dramatic and related to the level of milk output (Fig. 2.22).

The pregnant mammal fully anticipates impending lactation and lays down fat in order to lose it through milk. The young seal will grow 1.5 kg/day – putting farm livestock of any age or weight to shame – on milk containing 50 per cent fat produced by the dam, which eats nothing.

Figure 2.22 *Fat losses in dairy cows.*
(Unpublished results from the Langhill Project, University of Edinburgh)

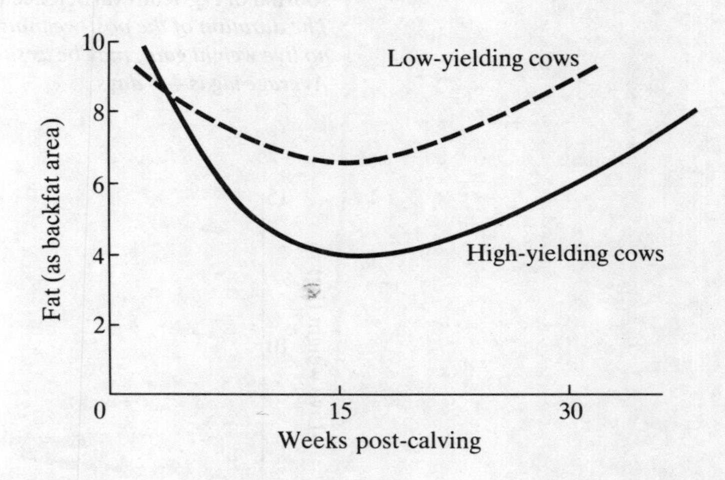

Daily losses of fat in lactation by dairy cows can be more than 1 kg/ day, while the underfed lactating sow can lose 0.5 kg. In terms of metabolic body weight, these amounts are similar at about 10 g/ kg metabolic body weight – the same rates of loss occur in lactating ewes.

Fat losses by young mammals after weaning relate to the extent of pre-weaning fat stores built up and the physiological need to maintain fat, even in the face of post-weaning nutritional stress. Thus fat little pigs will catabolize almost exclusively fat when deprived of adequate feed, whereas the less fat lamb and calf may also catabolize some small amounts of protein as well as fat. Weaned baby pigs preferentially lose subcutaneous fat to internal fat in the ratio of about 9:1, but the weaned seal, with a different view of its needs to combat the environment, will lose fat preferentially from the core rather than the blubber – the ratio being 6:4.

Immediately after weaning, pigs usually show weight stasis, live growth impulsion picking up about 1 week later (Fig. 2.23). The extent of this post-weaning growth check may vary from 3 to 14 days, depending upon both pre-weaning and post-weaning management, housing and nutrition.

It is apparent that weaning is a traumatic time for sucking animals. Pigs of 28 or 21 days of age on the day of weaning have body compositions of approximately 15 per cent protein and 15 per

Figure 2.23 *Growth of pigs weaned at 21 (broken line) and 28 (solid line) days of age. (From Whittemore, C. T., Aumaitre, A. and Williams, I. H. (1978).* Journal of Agricultural Science *(Cambridge)* **91***: 681)*
The duration of the post-weaning lag phase, in which piglets make little, or no live weight gain, may be as short as 2 days or as long as 2 weeks. Average lag is 4–8 days

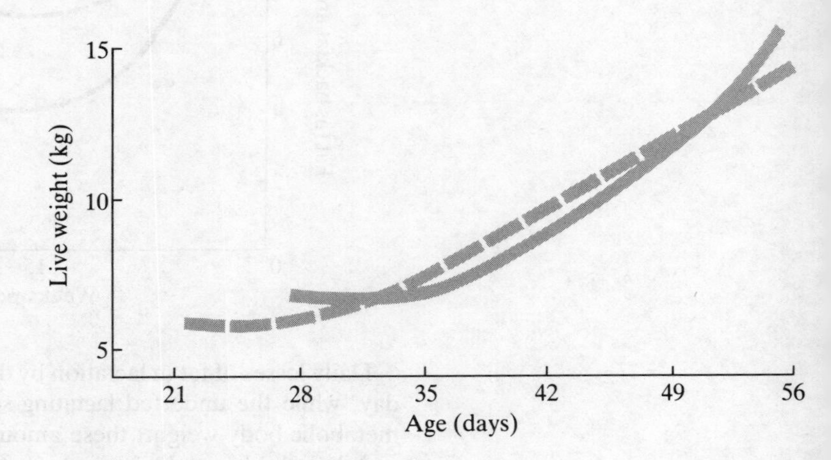

cent fat, while pigs 7 days after weaning have 15 per cent protein, but only 7 per cent fat; the percentage of fat in the total body being halved within 1 week. The apparent ability of these pigs to circumvent the laws of proportionality is due to the percentage of water in the empty body, which increases to counterbalance the fat.

Figure 2.24 shows data from four separate experiments. In experiments 1 and 2 pigs were weaned at 28 and 21 days of age respectively, and the compositions shown at these ages therefore relate to the sucking pig. In experiment 3 the pigs were weaned at 14 days of age and the composition given therefore pertains to 7 days post-weaning. It is apparent that weaning has not only brought about a great reduction in the percentage of fat, but fat losses have been compensated for on a proportional basis by the addition of water. At 55, 53 and 42 days of age, pigs in experiments 1, 2 and 3 had failed to make good their post-weaning

Figure 2.24 *Percentage composition of pigs. Pigs on experiments 1 and 2 at 28 and 21 days of age were taken at the point of weaning. Pigs on experiment 3 were weaned at 14 days of age.*
(From Whittemore, C. T., Aumaitre, A. and Williams, I. H. (1978) Journal of Agricultural Science *(Cambridge)* **91***: 681)*

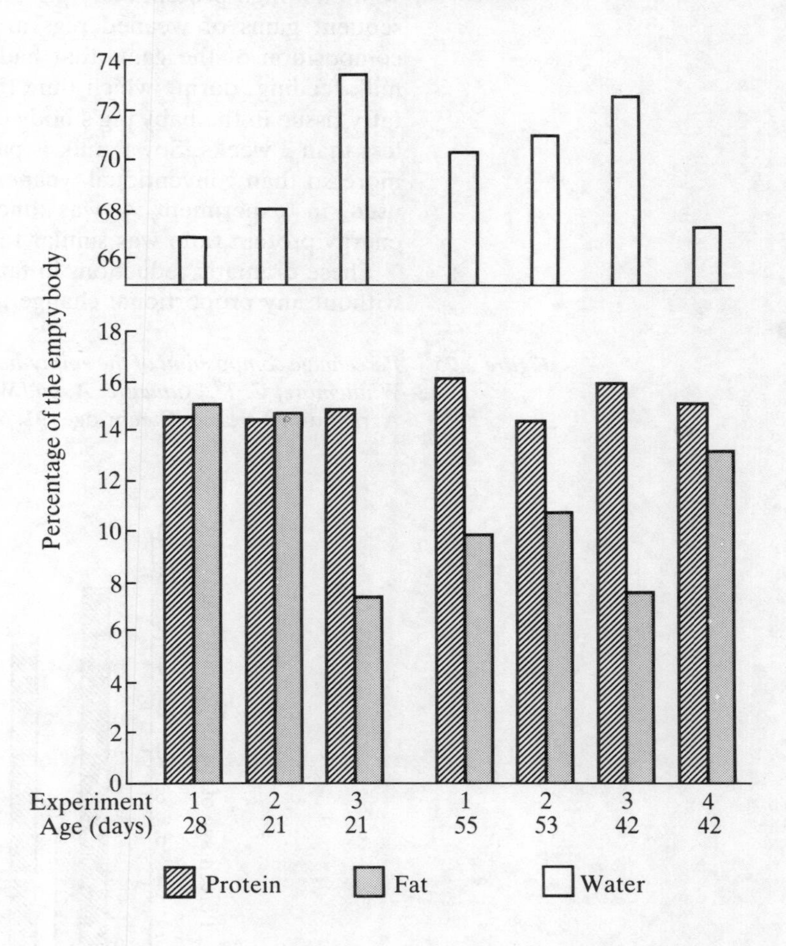

fatty tissue losses, but pigs on experiment 4 had done so. These latter pigs were given an unconventional diet with a high energy:protein ratio. Over the period of the experiments the composition of the post-weaning live growth in experiments 1, 2 and 3 was about 15 per cent protein and 7 per cent fat (Fig. 2.25), but in experiment 4 there was more fat than protein in the gain (about 15 per cent protein and 19 per cent fat). Despite the extent of fatty tissue loss in the period immediately after weaning, on conventional high-protein baby pig diets the composition of the subsequent gains of weaned pigs are much lower in fat than the composition of the gains that had previously been made during milk feeding, during which time the proportional composition of fatty tissue in the baby pig's body changes from 1 to 15 per cent in less than 3 weeks. Sows' milk is particularly rich in energy (much more so than conventional weaner diets). The post-weaning diet used in experiment 4 was unconventional inasmuch as its energy:protein ratio was similar to that of sows' milk.

These dramatic reductions in fatty tissue appear to be achieved without any proportional change in protein (Fig. 2.26). Thus the

Figure 2.25 *Percentage composition of the empty-body gain of pigs. (From Whittemore, C. T., Aumaitre, A. and Williams, I. H. (1978)* Journal of Agricultural Science *(*Cambridge*)* **91***: 681)*

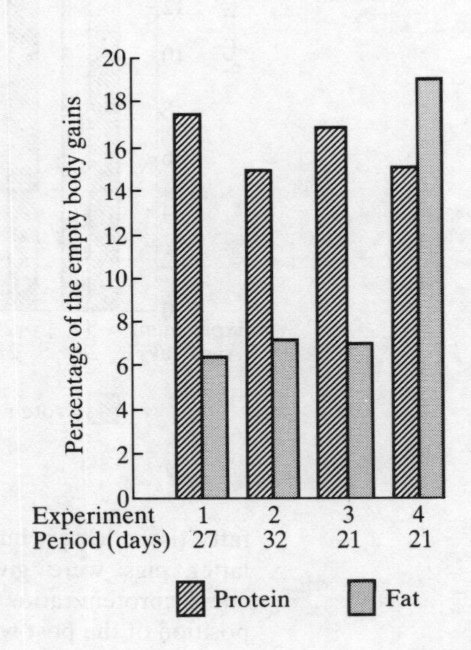

Figure 2.26 *Protein content of young pigs. Open symbols relate to unweaned animals. (From Whittemore, C. T., Aumaitre, A. and Williams, I. H. (1978)* Journal of Agricultural Science *(Cambridge)* **91***: 681)*

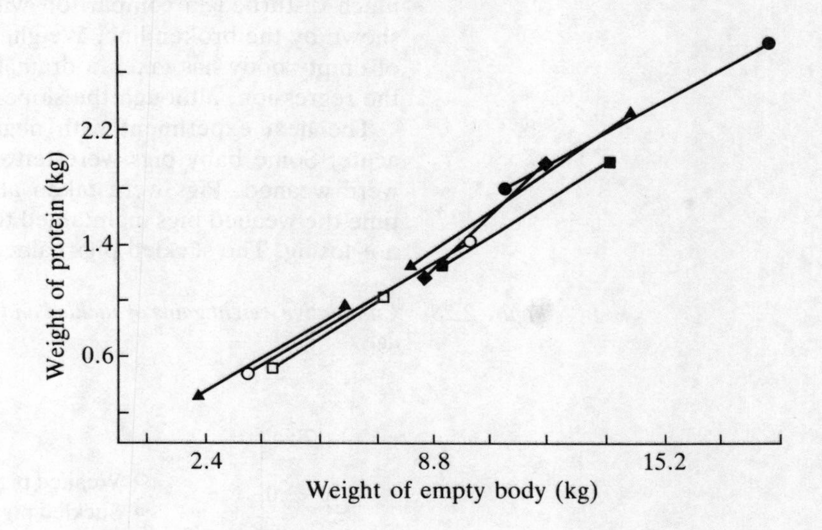

Figure 2.27 *Lipid content of young pigs. Open symbols relate to unweaned animals. (From Whittemore, C. T., Aumaitre, A. and Williams, I. H. (1978)* Journal of Agricultural Science *(Cambridge)* **91***: 681)*

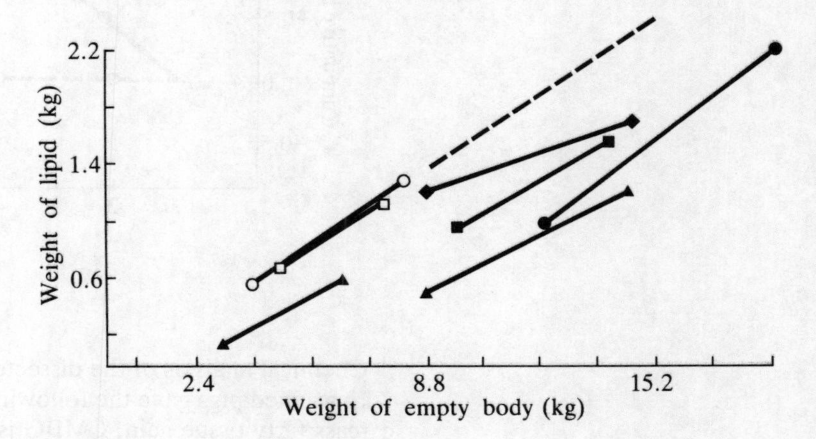

weight of protein as a function of empty body weight maintains the same relationship before and after weaning.

In the case of lipid, the picture is quite different (Fig. 2.27) and much disturbed in comparison with the unweaned extrapolation shown by the broken line. Weight of lipid as a function of weight of empty body has taken a dramatic fall in the constant term for the regression, although the slope appears little changed.

The next experiment with negative growth was rather more acute. Some baby pigs were left sucking the dam, whilst others were weaned. Pigs were taken at 2-day intervals, during which time the weaned pigs maintained their live weight, neither gaining nor losing. The suckled pigs gained around 300 g/day (Fig. 2.28).

Figure 2.28 *Cumulative weight gains of suckled and weaned pigs from 21 to 28 days of age*

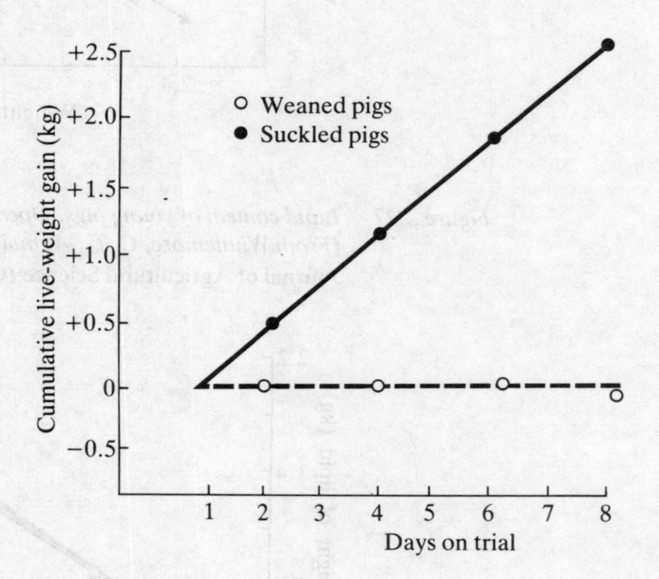

Chemical analysis of the dissected body components of these 6–7 kg weaned pigs gave the following relationships, where CFTG is carcass fatty tissue gain, CMBG is carcass muscle plus bone gain, WG is water gain, LG is lipid gain and PG is protein gain:

$$WG(g/day) = 0.16\ CFTG + 14.8$$
$$LG(g/day) = 0.78\ CFTG - 17.7$$
$$PG(g/day) = 0.06\ CFTG + 2.9$$
$$WG(g/day) = 0.70\ CMBG + 9.2$$
$$LG(g/day) = 0.08\ CMBG - 3.1$$
$$PG(g/day) = 0.21\ CMBG - 8.7$$

Figure 2.29 ***Relationships between total gain and the components of the gain. (Calculated from Whittemore, C. T., Taylor, H. M. Henderson, R., Wood, J. D. and Brock, D. C. (1981)*** Animal Production **32***: 203)*

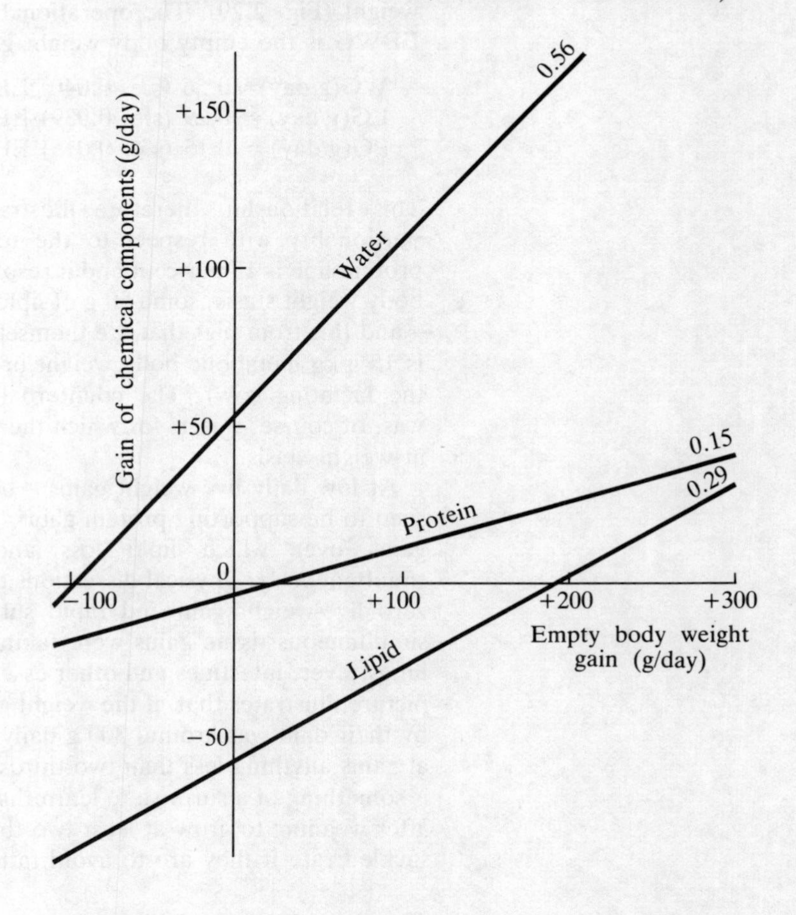

These relationships show the carcass fatty tissue gains to have been 16 per cent water, 78 per cent lipid and 6 per cent protein, while the gains of carcass muscle plus bone were 70 per cent water, 8 per cent lipid and 21 per cent protein. Zero gains of fatty tissue appear to be associated with gains of about 15 g of water and 3 g of protein.

There was sufficient variation in the total live-weight gains for regression relationships to be drawn up between the gains of the chemical components and the gains of the total empty body weight (Fig. 2.29). The operational regression equations, where EBWG is the empty body weight gain, were:

$$WG(g/day) = 0.56\ (s.e.\ 0.040)\ EBWG + 53\ (s.e.\ 6.8)$$
$$LG(g/day) = 0.29\ (s.e.\ 0.039)\ EBWG - 56\ (s.e.\ 6.7)$$
$$PG(g/day) = 0.15\ (s.e.\ 0.015)\ EBWG - 4\ (s.e.\ 2.5)$$

This relationship therefore illustrates the compositional proportionality with respect to the total gain. The slope of the protein line is 15 per cent and it resolutely passes near to zero. At body weight stasis, some 50 g of lipid are lost from the body daily – and this from pigs that are themselves only 5 kg in weight (that is 15 g/kg metabolic body weight or about 50% more than from the lactating cow). The counterbalancing chemical component was, of course, water, for which there were positive gains of 50 g at weight stasis.

At low daily live-weight gains – below 200 g – lipid losses are seen to be supporting protein gains, and there is quite a range of gains over which lipid loss and protein retention occur simultaneously. Physical dissections of the tissues showed that, at zero live-weight gain and rapid subcutaneous fatty tissue loss, simultaneous tissue gains were taking place mainly in the heart, lungs, liver, intestines and other essential body components. This picture illustrates that, if the weight gain of the pigs being suckled by their dam was around 300 g daily, fat losses may be expected at gains anything less than two-thirds of an apparent optimum. It is something of a surprise to learn that little pigs, anyway, require after weaning to grow at least two-thirds as fast as the unweaned suckled rate if they are to avoid fatty tissue breakdown.

Compensatory or catch-up growth

Slow growth and negative growth lead to the proposition of compensatory growth. If an animal's growth is retarded by nutritional deprivation at one stage in life, then counterbalancing adjustments take place at a later stage with enhanced efficiency during the adjustment phase, such that overall benefits accrue. The vital question is as to whether there is compensation in lean tissue gains in a way that a previously limiting boundary is now exceeded (Fig. 2.30). Such a possibility appears remote. There are a number of ways of appearing to prove compensatory growth during re-feeding after a phase of growth retardation. One is to fail to measure true weight gains and to assume that gut-fill is body- proper. Another is to wait sufficiently long that the control-group slows down as maturity is approached while the re-fed group, being lighter at the same age and therefore not slowing down, appear to catch up. The third and perhaps most popular way is to ensure that control animals, although free-fed, are – in fact – not maximizing their lean gains. Then, compensating animals may grow lean faster than controls when maximization of lean gains is enabled under re-feeding circumstances. That the control is performing at maximum and not

Figure 2.30 A working analogy for compensation

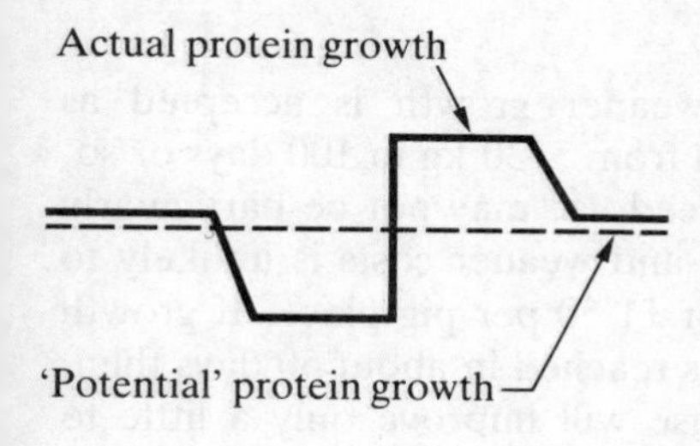

Figure 2.31 Growth of previously restricted pigs

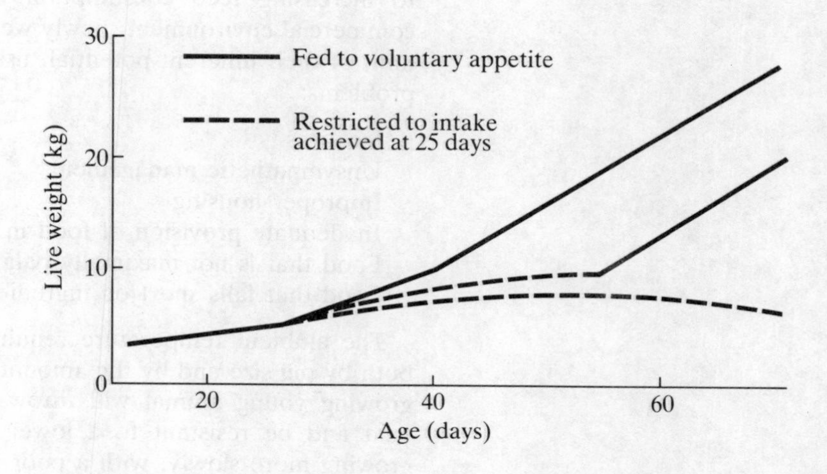

merely at average expectation must be an obligatory requirement for a compensatory growth experiment.

It is probable – but not certain – that lean growth once lost cannot be regained. For example, it appears from Fig. 2.31 that with pigs the growth of previously restricted animals is none other than parallel to the growth of unrestricted controls. Compensatory growth is still a great puzzle and the basis of active research at Edinburgh.

Some practical considerations for the care of newly weaned pigs

On many pig units, retarded weaner growth is accepted as normal. If weaned pigs are grown from 5–30 kg in 100 days or so, then, although the efficiency of feed use may not be particularly poor (0.4), the margin over feed and weaner costs is unlikely to be more than 50 pence per pig or £1.50 per pig place. If growth rate is improved such that 30 kg is reached in about 50 days then, although the efficiency of feed use will improve only a little to 0.5, the margin over feed and weaner costs will rise to more than £3 per pig and £20 per pig place. Further, the faster growing pigs require half the housing space and time, and therefore command only half the fixed costs and other (non-feed) variable costs.

Fast growth is achieved by obtaining a high level of feed intake. The relationship between daily gains of lean and fat and feed supply are such that there is a linear response of lean growth to increasing feed consumption by pigs of this age. In the commercial environment, newly weaned pigs fail to eat up to the limit of their inherent potential, usually because of the following problems:

- Sickness
- Unsympathetic management
- Improper housing
- Inadequate provision of food in inadequate feed receptacles
- Food that is not maximally palatable
- Food that falls short on ingredient quality

The ambient temperature required by young pigs is affected both by pig size and by the amount of food consumed. A rapidly growing young animal will throw off considerable amounts of heat and be resistant to a lower temperature than an animal growing more slowly, with a poor appetite and low feed intake. Newly weaned pigs require an air temperature of between 25 and 30°C, the higher number relating to those animals of lower feed intake and growth rate. Slow growth militates against the animal being able to deal with the challenge of a low or fluctuating temperature and further predisposes the animal to invasion by disease organisms.

Weaner growth may often be retarded because the pigs are too densely stocked. Over-stocking reduces feed intake and increases the likely spread of disease organisms. Keeping the numbers of pigs per pen at a high level is seen by some as an appropriate way to maintain a high air temperature and to solve problems of a shortage of housing space. The contrary, and more logical, view is that at reduced stocking rates the pigs will grow faster and therefore throw off more heat, while increased growth rate will enhance throughput and reduce pressure on accommodation.

Diets given in the post-weaning phase of life require to be nutritionally dense and highly digestible. But most are important of all is that the diet must be eaten. In the early stages of solid feeding the amount of food consumed is a more important consideration than the chemical compositional quality of the diet presented. Ingredients require to be of high acceptability to piglets, such as cooked cereals, milk products, animal proteins and high grade fats. The standard of the materials needs to be at the 'human grade' level and quite above all suspicion of contamination or unwholesomeness.

The journey time between food and water has to be short if maximum food intake is to be achieved; watering facilities therefore require to be close at hand to the dry-feed trough. To avoid the food becoming stale and unpalatable it needs to be provided in the trough at frequent intervals (2 or 3 times daily), and food should not be allowed to remain in the trough for more than 12 hours. Often trough space is limiting in many commercial weaner pens, the piglets not being able to get to the trough for the presence there of more aggressive companions.

Epilogue

The propositions developed here are not wholly in accord with established wisdom, but are important to the efficiency of meat production. The general assumptions relating to exponential growth in early life lead to expectations of maximum productivity at about half of mature size, whereas this may be obtained much earlier in life to the benefit of both man and animal. The causes of lost growth potential in young animals are ill health or inadequate nutrition. Both these shortcomings may be resolved and there is now ample evidence to show how growth impulsion in early life is primarily nutritionally motivated. Fatty tissue deposition during normal growth relates to the interactions of feed intake, feed quality and lean tissue growth rate. While the need for fatty tissues in meat has now diminished, fat catabolism plays an important role in providing energy in times of nutritional stress, such as after abrupt weaning and during lactation.

Chapter 3 Pig Carcass Quality

Introduction

In the recent past, the pig meat market has been 'mass and low price'. Indeed, fat reduction, now so happily achieved, was instigated not by consumer discrimination against fat, but by the economics of pig production and of pig-product processing. The market now is divided into two clear elements: first, the flexibility of pig meat as a substrate for a huge variety of manufactured meat products will ensure a continuing need for low-cost meat; but the second element is of ever increasing importance – this is for a meat of high eating quality. The consumer is demanding quality pork and bacon products as first choice for a main meal. This market needs carefully prepared and butchered cuts, a low level of subcutaneous fat, an adequate level of intramuscular fat and a large mass of lean tissue giving a high perception of eating pleasure; that is to say, juicy, tender and tasty – and avoiding all suggestion of dryness, paleness, taint or toughness.

Producers have often been reluctant to accept the restraints asked of them by a meat trade trying to maintain product quality. Conversely, the trade has sometimes been slow to allow producers to profit from the benefits of new techniques and streamlined production practices. These sorts of problems will remain with the industry while it continues to be based upon adversarial trading at each interface of the production chain. The benefits of integration between breeder, producer, feed compounder, slaughterer, meat trader and retailer are self-evident, and essential to the future of the industry. The producer needs better lines of communication with his market. If that communication is to generate information that is useful, both parties – not one party – must profit. An environment of opportunistic trading between the various factions within the industry, and an adversarial view where each believes it is exploited at the hand of the other, is counterproductive.

The business of producing meat from pigs requires that the producer has a clear view of his target end-product, has an adequate definition of that target and has a means of manipulating the production system to achieve the stated goal.

Usually, but not invariably, the pig buyer and meat packer will have some means of grading meat pigs received from producers; payment for the pigs will be related to the achievement of the grading standards. This does not mean, however, that the grading target set by the producer to optimize his business should necessarily be the same as the grading standards set by the buyer. A market preferring and paying premium rates for a high standard of blockiness in the carcass (as would be attained with Piétrain or Belgian Landrace-type pigs) may be best exploited not by attaining all pigs of premium grade from the use of purebred animals, but by a lower percentage in the premium grade from a cross with Large White animals, which would be more prolific, faster growing and less prone to stress. A market preferring and paying premium rates for pigs of less than 14 mm backfat depth at the P2 site (see Fig. 3.1) may often be best exploited by achieving less than 80 per cent top grade; for example, in circumstances where mixed groups of pigs require the castrated males to be grown slowly in order to achieve adequate leanness and where this causes the naturally leaner females to be unnecessarily held back. Sometimes the value of a grade premium can be less than the cost of its attainment.

Discrimination by a meat buyer against entire male pigs may be

Figure 3.1 Position of introscope probe measurements P1 (45) P2 (65) and P3 (80), in mm from the mid-line. Note the two fat layers and the third around the P3 region. Eye muscles can vary in shape – many are more oval than shown here.

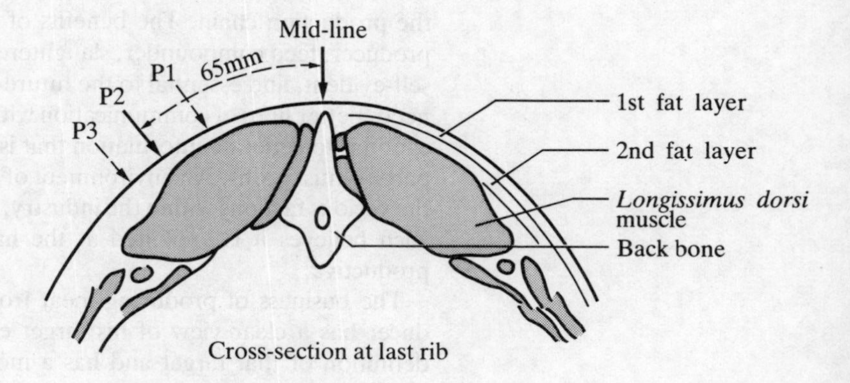

resolved by the producer castrating his animals, or alternatively by acceptance of a price penalty; optimum tactics will depend upon cost benefit analysis and the ability of the producer to encourage high feed intakes and rapid growth.

Pig producing businesses must therefore identify their target grading standards in the light of grading and payment schemes set up by meat buyers and processors; but targets for producers need not necessarily be identical with standards set by meat buyers. Producers should only pursue set targets after the optimum grading profile for any particular production unit in any particular trading environment has been decided.

Consumption patterns

The report of the Committee on Medical Aspects of Food Policy relating to *Diet and Cardiovascular Disease** (COMA report) makes specific mention of the need for a reduction in the level of fats – especially animal fats – in the human diet. Some 30–50 per cent of the total energy in the diets of many people in the advanced western world comes from fat of vegetable or animal origin. There has been a dramatic reduction of fat in pig meat, down from about 30 per cent of total carcass in the 1950s to about 15 per cent or less in a 90-kg bacon carcass in 1985. Backfat measurements have been reduced by genetics, nutrition and feeding management by about 0.5 mm annually. For top-grade bacon, backfat depth targets at the P2 site for pigs of 85 kg live weight are now often as low as 12 mm. On this basis it may be assumed that, on health grounds alone, the consumption of pig meat and pig products will increase.

In terms of value for money, pig meat is a better buy than beef or lamb, and rivals chicken and turkey (depending upon the joint in question and the assumed meat yield from a whole chicken roast). Pig meat is highly versatile, being used for fresh meat joints, cured joints, bacon, cooked meats (fresh and cured), processed products, patés, barbecue ribs, and many specialized lines of sausages and delicatessen items.

Most advanced European countries eat about 70–90 kg of total meat per person per year; this is about 30 per cent less than in the USA. In the UK much less pig meat is eaten than in Denmark, the Federal Republic of Germany or the Netherlands and it is also a lower proportion of total meat consumption (Table 3.1). In the UK the consumption of imported cured pig meat and pig meat

*(1984). HMSO, London.

products accounts for 30–40 per cent of total consumption and comes from a variety of sources, including Denmark, the Netherlands, Poland and Germany.

Table 3.1 Production and Consumption Trends. (*From* EAAP long range study on the future of pig production in Europe)

	Population (millions)	Consumption of pig meat (kg per head)			Self sufficiency (%)		
		1960	1980	2000	1960	1980	2000
Denmark	5	35	45	50	375	370	375
Federal Republic of Germany	56	35	60	70	90	90	90
Netherlands	12	25	40	50	150	225	300
UK	54	20	25	30	55	65	70

- Consumption of pig meat is gradually increasing
- Of total meat currently consumed, pig meat represents
  64% in Denmark
  62% in the Federal Republic of Germany
  55% in the Netherlands
  36% in the UK
- It is possible that in some countries, such as the Federal Republic of Germany, failing new products, consumption of pig meat is nearing saturation point.
- Exports from the Netherlands are increasing
- UK is far from self-sufficient (especially for bacon products) even though it has a conservative level of consumption

The efficiency of transfer of animal feed into pig meat is already high (Fig. 3.2) and continues to improve with new production technology. Increasing the product range available to UK consumers is probably the most appropriate way now open for the further expansion of the pig meat sector.

Aspects of grading standards

Grading schemes usually contain a range of standards. In some the standard (such as level of fatness) may be the prime controller of value and payment received by the producer, whilst in others the standard (such as quality of fat) may be of considerable importance to carcass quality but yet not play a functioning part in the payment schedule. Usually payment is on weight of pig (£ per kg dead weight) adjusted for quality assessment. Sometimes a flat-rate payment is made regardless of quality or grade attainment, while in countries such as Canada, payment is a 'flat rate' per kilogram of lean meat produced.

Figure 3.2 *Efficiency of transformation of feed to meat.* (*From* EAAP long range study on the future of pig production in Europe)

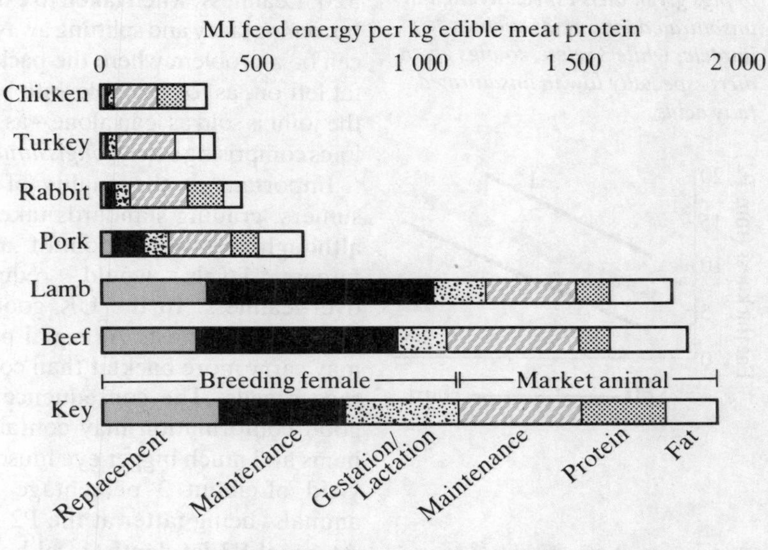

- Pigs compare favourably as meat producers
- Pig meat could be produced even more efficiently by:
  - increasing rate of growth (reduction of ▨)
  - increasing numbers reared/sow (reduction of ■)
  - decreasing production of fat (reduction of □)

In some markets conformation is a functional aspect of the grading standard. Shape is almost entirely dependent upon breed and not much open to nutritional manipulation other than through the creation of over-fatness. Most aspects of meat quality, other than amount of fat, are also not open to nutritional control. For example, muscle quality (as particularized by PSE and DFD*) primarily results from breed and physical treatment around the time of slaughter. Soft fat in pigs is positively associated, in any breed or sex, with leanness itself, but there may also be a tendency for entire males to have slightly softer fat than females, even at the same degree of fatness. Unsaturated fatty acids in the diet,

* Pale, soft and exudative (white, flabby and wet); dark, firm and dry (unappetising and tough).

Figure 3.3 Influence of fat thickness upon fat quality: 1 relates to pigs given diets especially high in unsaturated fatty acids such as linoleic, while 3 relates to pigs given diets especially low in unsaturated fatty acids.

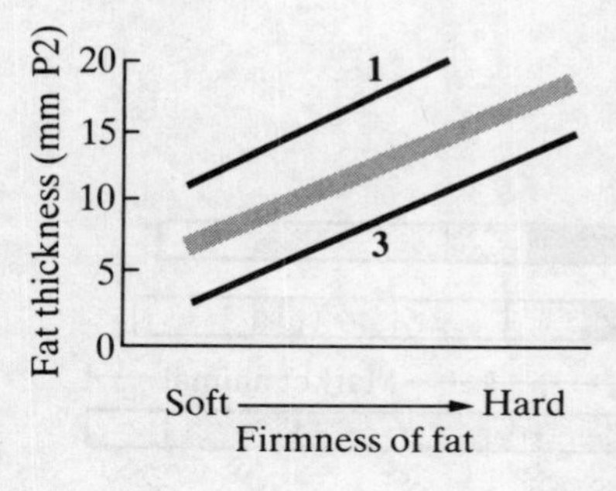

Figure 3.4 As fat depth reduces, the tendency for fat to split increases.

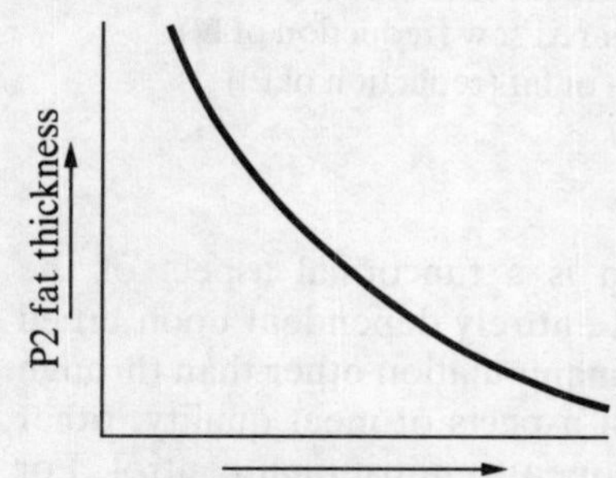

especially substitution of stearic (C18:0) with linoleic (C18:2), may influence the type of fat deposited in the body, rendering it softer (Fig. 3.3). Leanness, when taken to extremes in pigs, can also lead to the fat becoming lacey and splitting away from the lean (Fig. 3.4). Splitting fat can be a problem where the packer wishes to prepare cuts with some fat left on, as for example with bacon, but is less of a problem where the joint is sold as lean alone – as would be the case for continental pig loins comprising solely *longissimus dorsi* (eye) muscle.

Important as the quality of the lean and fat is to meat consumers, grading standards take little account of them at present, although the imposition of minimum, rather than maximum, fatness levels would reduce problems associated with over-leanness. In the UK good carcass shape may even be discriminated against. At equal percentage lean, the blocky breeds may carry more backfat than conventional Large White and Landrace breeds. The consequence is that, although some breeds of good conformation may contain more lean meat (having deeper hams and much bigger eye muscles, less bone, an improved carcass yield of about 3 percentage units, and smaller heads), these animals, being fatter at the P2 site, are liable to be down-graded. At equal P2 fat depth, total body fat may be 5 per cent less. The Piétrain has been estimated to have some 4 percentage units more lean meat than the Large White (61*v*.57%) and to have a considerably more favourable lean:bone ratio (6:1*v*.5:1).

Most grading standards and payment schedules relate to fatness grades (usually as backfat at the P2 site; Fig. 3.1) within a given weight band. For example, one scheme may limit head-on carcass dead weight to between 60 and 75 kg, and pay premium price for pigs of less than 14 mm P2 and impose a maximum price penalty for pigs of more than 18 mm. Equivalent schemes might pertain for pork carcasses of below 60 kg carcass weight, and cutter and heavy pig carcasses of above 75 kg. There may be additional requirements, perhaps that the length of the carcass be greater than 775 mm, or the fat measurements at shoulder, mid-back and loin each be contained within a certain limit. It follows from the natural growth pattern of the pig that lighter animals have a lower fat thickness; thus, as the potential value of the carcass increases with its weight, the likelihood of it being down-graded because of over-fatness is also increased. The consequence would be a diminished price paid for each and every kilogram of lean meat provided. This inconsistency would be ameliorated by a grading scheme that paid for the kilograms of lean meat yielded. This could be calculated from knowledge of the carcass weight and the P2 measurement (together with perhaps the breed, if of the blocky type). Some suggested predictions of percentage lean in the carcass side have been 63 − 0.51P2, and 61 − 0.52P2; while the average multiplier for P2 over a further sample of data sets was −0.58. Workers at the Meat and Livestock Commission have presented three equations for 47-, 72- and 93-kg carcasses. These were respectively: 60 − 0.73P2, 60 − 0.63P2 and 57 − 0.54P2.

The phenomenon of increased fatness with increased weight would also indicate that the correct slaughter weight for the fatter castrated males should be towards the lighter end of the market, whilst that for thin entire males should be towards the heavier end. It is germane to production tactics that, unless controlled, the natural tendency is for fatness to increase faster than body weight, thus fatness accelerates disproportionately rapidly. This is well illustrated by the prediction of subcutaneous fat (kg) (Y) from carcass weight (X) in one population of pigs: $Y = 0.0002X^{2.54}$.

Strict weight limits for any particular grading scheme are often not needed in effectively integrated producer/retailer organizations: many buyers will take in pigs over a range of, say, 50–80 kg carcass weight, but nonetheless do this through the medium of not one, but perhaps three, separate grading schedules, each penalizing producers for underweights and overweights. Achievement of maximum allowable weight within any one scheme, consistent with an adequately low depth of backfat and an adequately low number of outgrades due to overweight, is clearly an appropriate producer target.

Individual grading schemes may contain many or few criteria. The more criteria, the greater the likelihood of an acceptable carcass on one criterion being found unacceptable on another. Some example criteria are given in Table 3.2. Standards of over-fatness (say a maximum of 14 mm P2) may come to be matched by standards of under-fatness. Minimum fat levels (about 8 mm P2) are required for the eating quality of the lean meat, as well as to help maintain the quality of the fat itself. Given modern ultrasonic techniques for measurement of carcass fatness in the live animal, pigs could, if required, be sent off to slaughter at a given target fatness of 10 mm P2. Targeting to sale within a given weight range can be antagonistic to targeting to sale within a given backfat depth range; the narrower these ranges become, the more difficult it is to meet both simultaneously.

Table 3.2 Some example grading criteria for pig carcasses

	Criteria	Rate of usage
Carcass dead weight	'Pork' $<$ 60 kg	High in UK
	'Bacon' 60–80 kg	High in UK
	'Heavy' $>$ 80 kg	High in UK
Backfat depth	P2 or P1 + P3	High in UK
	Mid-line at	
	Mid back	Medium in UK
	Loin	Medium in UK
	Shoulder	Medium in UK
Length	Length of warm, hung, carcass	Medium in UK
Shape	Hams	Low in UK, high in Europe
	Eye-muscle shape and size	Low in UK, high in Europe
	Distribution of joints	Low
Sex	Entire, female, castrate	Medium in UK, medium in Europe
Muscle quality	PSE, DFD	Low
Fat quality	Wetness, firmness, tendency to split	Low
Eating quality	Taint	Low
	Flavour, juiciness, texture	Low

Fatness is considered as normally distributed through the pig population. However, as average fatness reduces, so the possibility of normal distribution lessens and that for skewness increases. With a skewed distribution, the likelihood of over-lean pigs rises disproportionately. Given the perceived biological distributions for fatness and for weight at any given age, it is difficult to see how producers could supply pigs to an ever decreasing window size of weight, fatness and age. The benefits to the producer of sale for slaughter at pre-selected calendar dates may override the benefits of closely meeting arbitrary standards for weight and fatness.

Pig meat quality

In comparison with sheep meat and beef, there is much less effect of the anatomical position of the joint on the quality of pig meat. However, loins (*longissimus dorsi* muscle) are particularly appropriate for grilling or frying fresh, while the ham is a good shape for roasting. New cutting techniques currently being put forward by the MLC will do much to increase the product range and consumption of pig meat. The MLC method trims the carcass to a low-fat or no-fat state, uses natural muscles as the basis for jointing, and produces a range of steaks and rolls that well suit current demands for lean meat presented in a convenient way for handling and cooking (Fig. 3.5).

Figure 3.5 *'Providing the lean alternative with pork': taken from a Meat and Livestock Commission publication describing new cutting techniques.*

While P2 backfat depth is the major criterion of assessment of quality in pig carcasses as far as producer payment is concerned, it is much less important at the point of consumption – particularly if the joints are trimmed and prepared prior to sale.

Pig meat quality is usually defined in terms of the colour and water-holding capacity of the lean, the firmness of the fat and the liability of the fat to split away from the lean when bacon is sliced, packed or cooked.

Dry, firm and dark (DFD) muscle is relatively infrequently seen and may be attributed, quite safely, to poor pre-slaughter handling.

Pale, soft and exudative (PSE) muscle is 'seen' – albeit usually at a low and little-problem level – in about 15% of carcasses in the UK and is increasing. It is often, but not always, sufficiently severe as to make the meat of reduced acceptability for curing and processing, and sometimes possibly also for fresh meat sales.

PSE seems to be related to two causes: one is mishandling and pig stress, and the other is the genetic make-up of the pig. These interact such that susceptible pigs, when stressed, will show a very high frequency of PSE meat, while if unstressed the rate of PSE can be halved, even in fully susceptible animals. PSE is closely related to the presence in the genotype of the halothane gene and the variable influence of different abattoirs upon strains of differing susceptibility is shown in Fig. 3.6.

Figure 3.6 Influence of the presence of the 'halothane gene' upon PSE: nn is halothane positive, NN is halothane negative, and Nn is the crossbred. Responses are shown for two factories: 1, with poor handling facilities and 2, with good handling facilities.

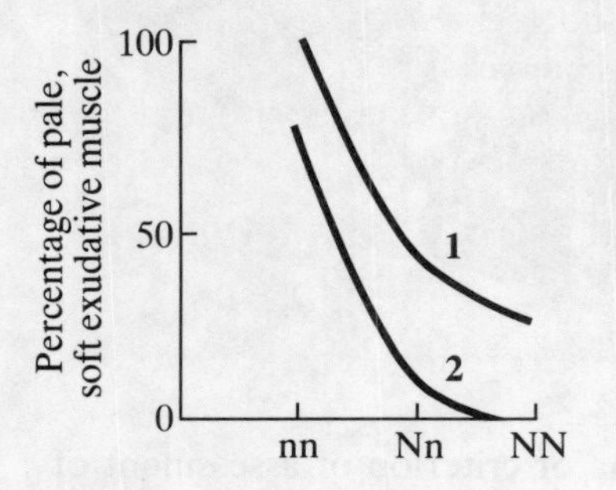

Eating quality of lean is often said to be positively related to fatness, and certainly a small minimum quantity of fat seems to be needed for succulence in beef and sheep meat. With pig meat, however, most fat is subcutaneous, intramuscular fat only increasing from 1 per cent in lean pigs to about 3 per cent in fat pigs. In general, it is thought that best eating quality is achieved at between 8 and 16 mm of P2 backfat. At below 8 mm the quality of the lean falls, while above 16 mm the meat is too fatty. Over the range 2–4 per cent intramuscular fat it is quite difficult to show anything but the slightest improvement in tenderness, succulence and flavour. Fatness levels of greater than 4 per cent can begin to affect adversely eating enjoyment, while at less than 1.5 per cent of fat in the muscle it is possible that tenderness and succulence are similarly adversely affected. Such a low level of intramuscular fat may be expected to be associated with pigs of P2 backfat depths of less than 8 mm.

There is current, and possibly transient, interest in the use of the Duroc as a top-crossing sire (or part of one) for the production of high-quality eating pork. The Duroc is quite widely used in America and in some European countries as a 'third breed', that is, as the top-crossing male on to a Large White × Landrace sow. It has a reputation for imparting robustness to breeding females, but it is not so prolific as the Large White × Landrace hybrid. The

Duroc can be a meaty breed, although tending to fatness and also to slow growth (although some strains are less prone than others). But especially interesting is that there is a higher percentage of intramuscular (marbling) fat than is the case for conventional white breeds (2–4% rather than 1–2%), and this is said to give extra flavour, tenderness and succulence to fresh, but not cured or processed, meat. As an approximation the relationship between backfat and marbling fat is P2 (mm)/10 $\simeq$ percentage marbling fat. For the Duroc the divisor is 5. It is likely that, if required, similar characteristics could be bred for in other breeds not showing the disadvantages of the Duroc such as slower growth, greater subcutaneous fatness and poorer reproductive performance. Possible improvement to eating quality by enhancing intramuscular fat should, however, not be taken too far.

Shape and meatiness: the halothane gene

Some pig types are more meaty than others and have a characteristic blocky shape. These types invariably carry the 'halothane gene', so-called because the animals react badly to the anaesthetic halothane. This reaction may therefore be used as a marker for muscle shape and meatiness.

Halothane reactors are found at various levels amongst pig populations (Table 3.3) and differ between strains within breeds. Amongst breeds and crosses containing 40–70 per cent of the gene, its presence is associated with:

about 2.5 per cent more carcass lean,
about 1 per cent higher killing-out percentage,
about 10 mm reduced length of carcass,
about 50 per cent more carcasses liable to pale, soft and exudative muscle,
about one less pig weaned per litter,
a reduced appetite,
slower growth,
bigger eye-muscle (*longissimus dorso*) area, and
less bone.

Purebred Belgian Landrace and Piétrain pig types, and many strains of German Landrace, show greater benefits over the Large White pig types of up to 4 percentage units more lean (61 *v.* 57% in the carcass sides; Fig. 3.7), up to 3 percentage units better killing-out per cent (76 *v.* 73%), and less carcass bone and smaller heads (the lean:bone ratio being about 6:1 as against about 5:1).

Table 3.3 Incidence of the 'halothane gene', and associated meaty and blocky characteristics

	Occurrence (%)
Large White	≈ 0
Duroc Hampshire Norwegian Landrace Danish Landrace British Landrace Swedish Landrace	2–25 (variable between strains)
Dutch Landrace German Landrace	≈ 70 (much lower in some strains)
Piétrain Belgian Landrace	≈ 90

Figure 3.7 At any given fat depth the meat-type breeds of pigs tend to have a higher percentage of lean meat. Their shape is more blocky and muscular and the content of bone is less.

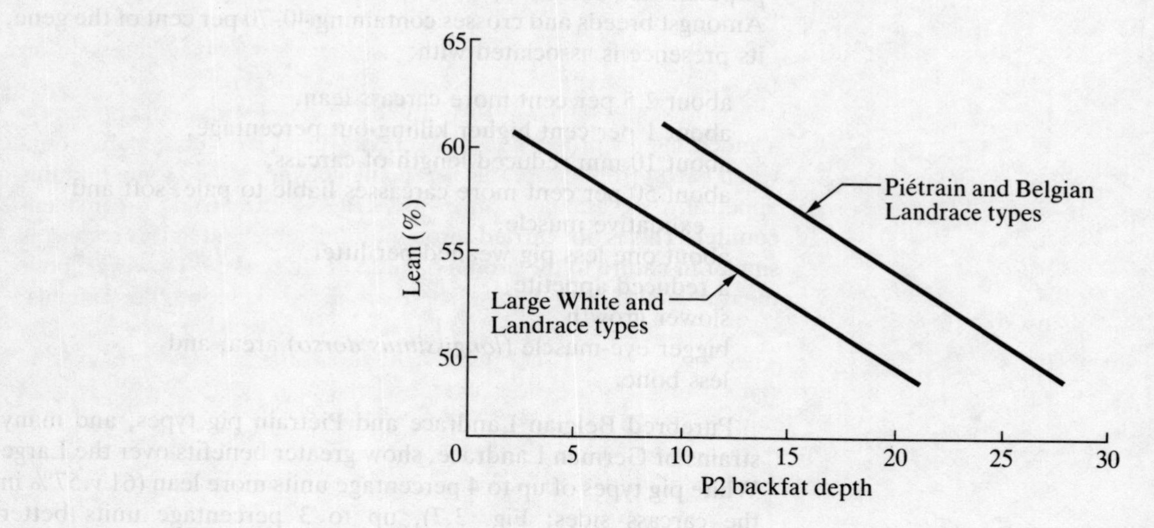

Other characteristics include smaller litters (10–20% less pigs weaned), slower growth rate (10–20%) and, of course, a tendency to drop dead when stressed (as, for example, when loaded into a lorry or when in abattoir lairage).

The characteristic shape of meaty pig strains is that they are short and blocky, and have well-filled rounded hams, the rear view being more Ω-shaped than n-shaped.

A well tried breeding strategy is to use a halothane negative female line (or cross) together with a halothane positive top- crossing meat sire line. The resultant heterozygotes carry 50 per cent of the benefits of the gene, but do not show the disadvantages of the porcine stress syndrome associated with homozygotes.

Meat from entire male pigs

Entire males are much less fat than castrates. At 90 kg, on average, an entire will have around 10–12 per cent of body lipid whereas a castrate will have 16–18 per cent. This attribute, together with faster and more efficient growth, makes the entire male an attractive proposition for the production of bacon and pork.

In many countries, no males are castrated and pigs are sold for slaughter at a range of live weights up to 100 kg. In other countries all males are castrated. In the UK about half of all male pigs are sold entire and that proportion is increasing; most of these go for fresh meat production, sold at less than 80 kg live weight, although an increasing number are now being cured for bacon, giving skin-on rashers with large eye muscles and low fat thickness.

Some people (about 5–10% of the UK population, but apparently a much higher proportion of Scandinavians) may detect an aroma when bacon from entire males is being cooked. While some find this appetizing, occasionally others may dislike the aroma on first encounter. This is not carried through to the table and in this respect is similar in nature to the aroma of lamb being cooked. However, boar meat may also carry taints. These are more likely in older animals, which are heavier (above 100 kg) or which have been grown slowly. The causes of taints are androstenone and, more seriously, skatole (which may on occasion also be found at lower concentrations in females).

In countless consumer trials over many years, taste panels have been unable to fault bacon and pork products from entire males. There is no evidence of differences in eating quality or acceptance, and often the greater eye muscle area and reduced fat content are found positively beneficial.

Entire males, being so much less fat than castrates, also suffer from the disadvantages of thinness; namely, having fat that tends to be wetter, softer and more liable to split. These may be remedied by limiting unsaturated fats in the diets and by increasing feed levels in order to avoid over-leanness. However, even at equal fatness, the fat of boars tends to be a little less firm than that of gilts. Butchers have complained that lean pigs are difficult to handle and cut, being floppier – this is mostly a characteristic of leanness, but entire males do seem more prone than females.

Current prejudice against entire males by some meat wholesalers seems to relate to problems with those few animals that are very lean and to equally few others that have been grown too slowly. Boars should only be purchased by meat abattoirs and processors under tight contractual arrangements from known producers who can guarantee fast, uninterrupted growth, and offer high-level feeding to the animals in a healthy production environment. Pigs failing to meet these minimum criteria should be diverted from the bacon and fresh pork market, and used elsewhere in the production line.

Problems of natural variation

In any production system there is variation; the less well managed the production process, the greater the variation. In Fig. 3.8, low-variation farm A will have no pigs at the far extremes of fatness and thinness, relatively few over-thin and over-fat pigs, and proportionately more at around the average. Common causes of increased variation are disease, poor feeding management and poor housing.

Producers will target to achieve required grade standards; should there be only one standard, say fatness, then success depends on the efficiency of the achievement of the standard and the smallness of the variation. Given a standard of fatness described by a maximum depth measurement of 14 mm, it is evident from Fig. 3.8 that farm A will have fewer outgrades than farm B, even though both farms have the same average fat depth.

Additional difficulties arise when more than one standard has to be met simultaneously. Often there are four main categories of carcass assessment:

Carcass weight
Carcass fatness
Carcass shape
Meat quality

Figure 3.8 Natural variation in fatness experienced on two farms, A and B.

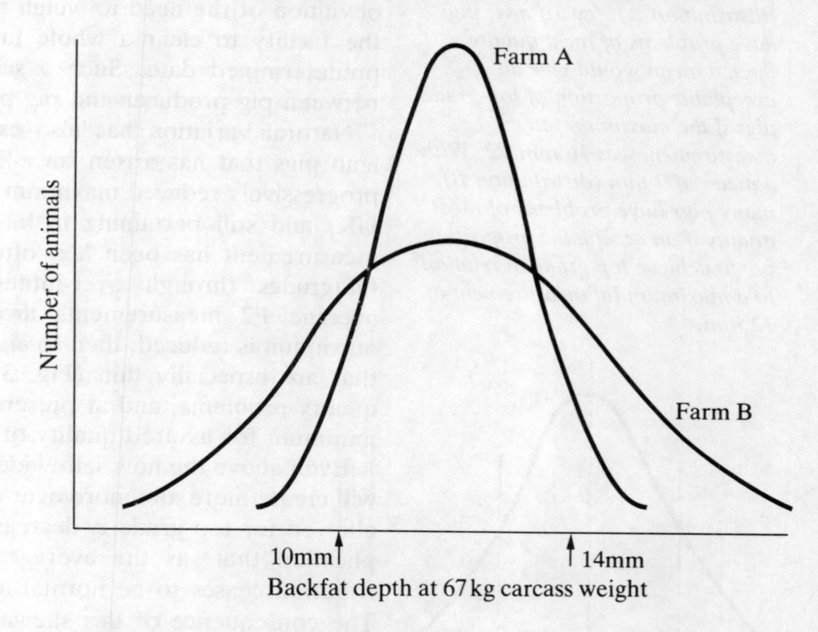

Shape and meat quality are primarily pre-determined by pig type, and pig and meat handling facilities at the abattoir. Fatness and carcass weight are both under the short-term control of the producer. Given a weight range of, say, 60–70 kg dead weight and a maximum fatness of 14 mm P2, there are more likely to be outgrades with carcasses at the heavy end of the weight range than at the light. This is because, as pigs get heavier, there is a natural tendency for them also to get fatter. While pigs are often individually weighed before despatch, they are rarely checked for fat depth. There is increasing pressure on fat and there is therefore some logic in giving consideration to measuring fat depth on individual pigs before despatch, in order to send fatter pigs off at the lower end of the allowed weight range and to ensure that a high proportion of pigs meet the fat depth target.

The alternative strategy is for pig processor and pig producer to accept that, overall, a range of carcass weights are required, and a range of fatnesses can be accepted within each carcass weight. Such an agreement may cost a little extra at the processor end of the business and this would have to be offset in producer prices.

Figure 3.9 With a mean of 13 mm (distribution A), few, if any, pigs have problems of meat quality. Such a mean would give an acceptable proportion of top-grade pigs if the maximum fat measurement was 16 mm P2. With a mean of 9 mm (distribution B), many pigs have problems of meat quality if an acceptable proportion are to achieve top grade in relation to a maximum fat measurement of 12 mm.

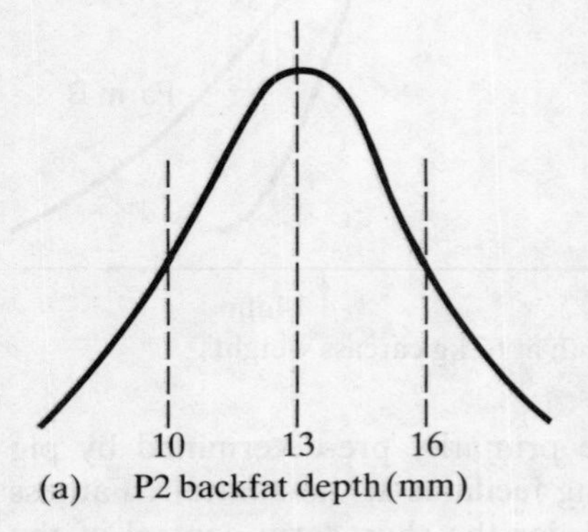

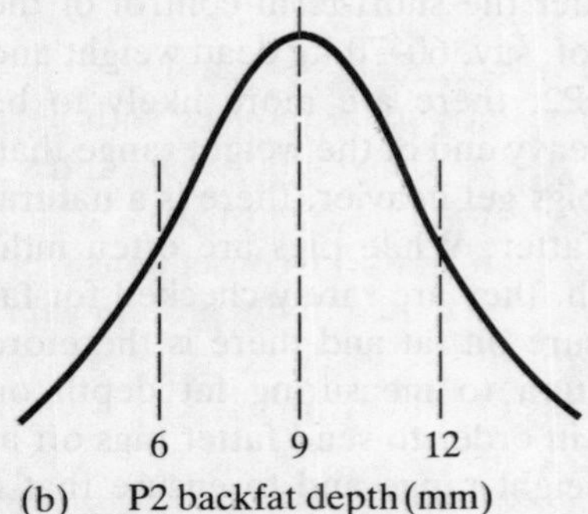

But the benefits to the producer could be legion – not least, obviation of the need to weigh and measure individual pigs, and the facility to clear a whole fattening house at one time at a predetermined date. Such a scheme requires close integration between pig producer and pig processor.

Natural variation has also exacerbated the problem of very lean pigs that has arisen since grade scheme requirements have progressively reduced maximum fat allowances. In the past in the UK, and still pertaining in many countries, the maximum fat measurement has been lax, often above 20 mm at the P2 site. Outgrades through over-fatness were therefore minimal at average P2 measurements around 15 mm. However, as the maximum is reduced, then so also is the number of pigs increased that are especially thin (Fig. 3.9). Over-leanness creates meat quality problems, and at present it appears that 8 mm P2 is the minimum for assured quality of lean and fat.

Even above the now self-evident problem that natural variation will create more and more over-lean pigs as the maximum fatness allowed for top grade is decreased, there is the additional complication that, as the average for fatness decreases, the distribution ceases to be normal and becomes skewed (Fig. 3.10). The consequence of this skewness is that proportionately more pigs fall on the lean side of the average, so heightening the difficulties.

It seems as if the window for fat thickness through which producers must fit their pigs is getting narrower. This will mean that pigs will need to be despatched upon a much better estimate of fatness than heretofore and probably somewhat more independent of weight. The production cost of manipulating the diet and management of individual pigs in order to produce all pigs of identical fatness at identical weight does not bear contemplation. Adjustment of carcass character after slaughter by the actions of the meat processor may often be cheaper than trying to produce carcasses ex-farm that are individually styled before entry to the abattoir to suit the exact needs of the pig processing line. None the less, reduction of the extent of variation by tight production management is a basic requirement of satisfying consumer needs and future patterns will certainly be to a distribution more akin to type A than type B depicted in Fig. 3.8.

Figure 3.10 *Showing that, as average backfat is reduced, the distribution changes from being normal to being skewed. The truncation of the distribution to remove 15–20% of pigs from the fat end of the scale will result in proportionately more pigs at the thin end if the distribution is skewed.*

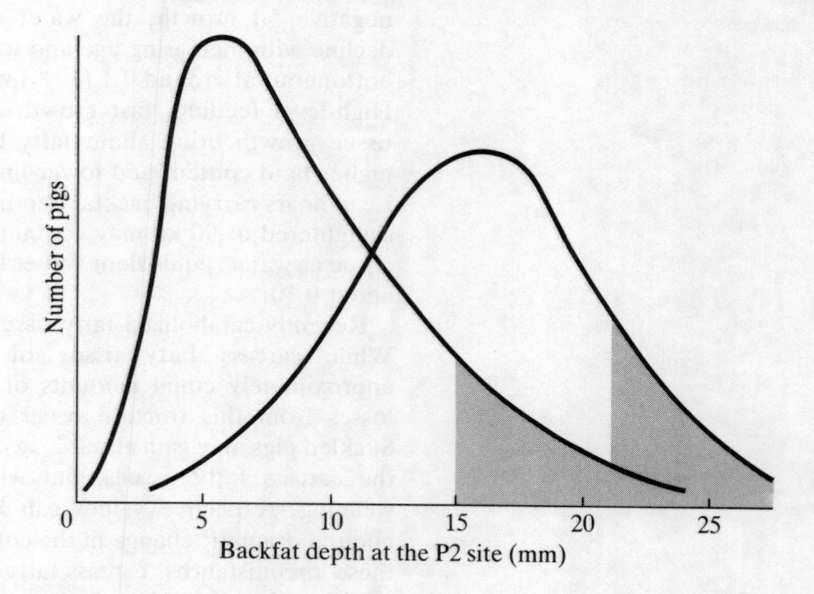

Body composition relationships in growing pigs

While variation in body composition is most clearly demonstrated by changes in the proportion of fat expressed as lipid:protein, there are also independent changes in contents of water, ash and carbohydrate. Water content of the body mass reduces with age and weight. Total body water may be approximated by the expression derived from Edinburgh data: 6 (total body protein)$^{0.8}$. The increase in the dry matter of the lean mass in consequence of advancing weight is also shown by: Lean gain = daily protein gain (5.1 −0.009 live weight), derived from the detailed chemical analysis of growing pigs by the Polish worker, Dr Maria Kotarbinska.

At equal fatness, boars have higher concentrations of water in backfat than castrates and lower concentrations of lipid; Dr J. D. Wood and his colleagues at the AFRC Institute of Food Research, Bristol Laboratory, found respective proportions for boars and castrates to be 0.16 and 0.11 for water, and 0.79 and 0.85 for lipid. Differences in the fatty acid composition of the tissue were slight.

The composition of the backfat of gilts and castrates appears to be similar. Young fatty tissue contains a high proportion of water (up to 0.8 after birth), which rapidly falls to about 0.2 (20%) at around 3–4 weeks of age. With the exception of the consequences of negative fat growth, the water content continues gradually to decline with increasing age and with increasing quantity of fat, to bottom out at around 0.1 (10%) water in the fatty carcass tissues. High-level feeding, fast growth and concomitant rates of fatty tissue growth bring about fatty tissues of lower water content, higher lipid content and lower linoleic (18:2) acid concentration. Lean boars carrying backfat depths of less than 10 mm at P2 when slaughtered at 90 kg may contain up to 0.25 of the carcass fatty tissue as water; equivalent values for gilts and castrates seem to be about 0.10.

Recently catabolized fatty tissue is open to invasion by water. While carcass fatty tissue of baby sucking pigs contains approximately equal amounts of lipid and water, post-weaning losses from this fraction occur concurrently with water gains. Suckled pigs may gain about 27 g of lipid and 20 g of water daily in the carcass fatty tissues, but within 2 days of the trauma of weaning, respective values can be −75 g and +17 g, bringing about a dramatic change in the composition of the fatty tissues. In these circumstances, carcass fatty tissue dry matter (CFT DM) is significantly related to the rate of empty body weight change (EBWC): CFT DM = 0.036 EBWC + 50. The consequential poor relationship between rate of live weight change and extent of fat catabolism has also been noted in lactating mammals.

Causes of variation in the extent of mineralization independent of the strong relationship with age/weight is not well documented. Spray and Widdowson in their classical early work chose to express mineral content of the body as a proportion of the fat-free body mass, and Kotarbinska suggests total body ash to be about 0.20 of total body protein. The equivalent coefficient estimated at Edinburgh is 0.21, while the ash content of gain may be estimated as 0.22 (protein gain). Weight of carcass dissected bone may be approximated from the total ash by multiplying by 2.5, or from the total protein mass by multiplying by 0.5. These latter exclude the bone content of the head. The level of dietary available mineral may contribute toward the ash in the bone, but the relevance of increased breaking strength to animal productivity is remote. However, the proportion of bone in the carcass is influenced by breed, such that breeds of the Piétrain/Belgian Landrace types have a more favourable lean:bone ratio (6:1, as compared with 5:1 for the Large White breed).

Liver carbohydrate is a major energy source in the neonate, when lipid content is extremely low. In later life, variation in body carbohydrate is not significant unless the pig is induced to consume high levels of sugar solution immediately prior to slaughter, when short-term liver weight can be doubled.

In pigs there can be important breed variation on body fatness independent of weight, age or nutrition. Prediction of proportions of carcass fat and of carcass lean from depth of subcutaneous backfat at the P2 site seems not to be consistently improved by the inclusion of live weight in the regressions. Recent estimates suggest that percentage lean in the carcass may be predicted from equations of the form $k - n\mathrm{P2}$, where P2 is millimetres of backfat 65 mm from the mid-line, and k and n are about 60 and 0.7 respectively. However, such equations suitable for Large White pigs may not be appropriate if the breed in question is of the blocky type, which can carry significantly more lean meat at equal P2 fatness. A more universal estimate of percentage lean in the carcass may therefore be of the form $k\mathrm{P2}^{-n}$, where n is about 0.21, and k indicates the degree of blockiness and ranges between 90 for the Large White and Landrace breeds, and 100 for the pure bred Piétrain/Belgian Landrace types.

Nutritional manipulation of carcass quality

Whilst for many pigs there are few or no grading standards and for others the standards include measures such as conformation, in the present UK environment pig grading standards can be considered as primarily dependent upon the thickness of backfat. Backfat depth is greatly influenced by breed and genetic merit; Large White pigs are leaner than Piétrain and over the past 20 years of selection in the Large White breed some 10 mm of backfat has been removed from pigs slaughtered at 90 kg. The primary determinant of grade is the quality or the strain of the pig in use; genetically fat pigs will tend to be always fat within the feasible range of nutritional and environmental variation. However, whilst the long-range strategy for fatness reduction and meeting grading standards must be genetic, the tactics with animals of any given genetic composition must depend upon the knowledge that fatness is greatly influenced by both quality and quantity of food. The major mechanism open to producers to manipulate grade and achieve grading standards is therefore through the control of the nutrition of the growing pig.

Classical growth analysis, such as proposed by Hammond, presents a view of waves of growth passing through the body as weight and age progresses, concluding with the fattening out process and tissue deposition in the later maturing body parts. Increasing fatness is proffered as an exponential function of advancing weight. It now appears that fatness itself, far from being related to age and weight, is instead a direct function of the level of nutrient supply in relation to the level of nutrient need; the latter being defined as the requirements for maintenance plus maximization of potential daily lean tissue growth rate. Modern strains of meat pigs may never contain more than the 150 g lipid per kilogram live weight, which is evident in extreme youth at 3 weeks of age when sucking. There is little ground for counting age or weight as significant contributors to body fatness in pigs grown for meat. In contrast, animals grown in nutritionally limiting circumstances may never fatten until maturity of lean body mass is obtained; and only in unimproved breeds maturing at small body size may the classical sequential process of fattening be observed. However, with fast growing improved breeds of larger mature body size, fattening (when and if needed) must be achieved in the course of the immature rapid growth phase and is dependent upon feed intake (voluntary or imposed) being in excess of the need of the animal to meet its aspirations to maximize daily protein accretion.

Protein Diets that do not adequately provide for the requirement of absolute amounts of protein fail to allow maximum lean tissue growth. Energy thus freed from the business of protein synthesis is diverted to fatty tissue growth. Excess protein, on the other hand, reduces dietary energy level. The energy yielded from protein by deamination is about half of the assumed digestible energy of protein. The effective energy value of a diet containing excess of protein will therefore fall as deamination rate rises, with a resultant diminution of energy available for fat deposition. Excess supply of total protein over the requirements for protein maintenance and protein growth will have the consequence of enhancing leanness. Increasing diet protein therefore first reduces fat in the carcass by diverting energy into lean tissue growth and away from fatty tissue growth as a frank protein deficiency is reduced. Secondly, further increments of protein above the requirement will continue to increase leanness by effectively reducing the available energy yielded from the diet and thus pre-empting the conversion of energy to fat. This second method, of reducing carcass fatness by the supply of excess protein, whilst explaining linear responses of percentage lean to dietary crude protein, can be expensive to execute and also bring about a reduction in pig growth rate. This latter is consequent not only upon a reduction in the rate of fat gains; if the general level of feeding is not sufficiently generous, then lean growth itself may be curtailed through an inadequate energy supply to drive the metabolic motors of protein anabolism. The role of protein concentration is illustrated in Fig. 3.11.

Figure 3.11 *Growth responses of lean and fat to increasintg dietary protein concentration.*

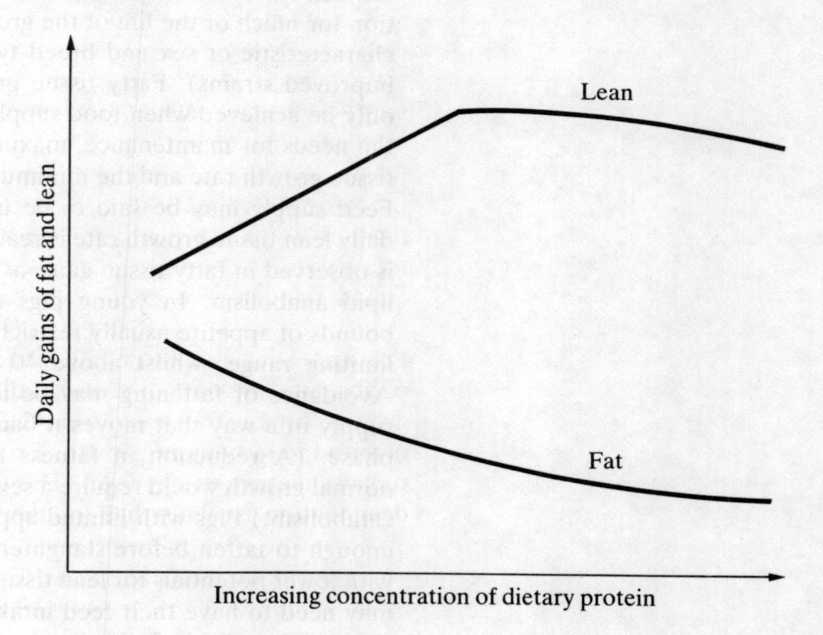

Level of feed (energy supplied) As modern improved strains of pigs may reach slaughter weight with less (sometimes considerably less) than 15 per cent body lipid, then it may be proposed that the balance of fat to lean in the growth should be rather constant over the whole of the growing period. The idea that, as the slaughter weight is approached, fattening should be induced by means of increasing the balance of energy to the protein is defunct (with the possible exception of breeding herd replacements); and therefore decreasing the level of dietary protein over the growing period is required only to the small extent needed to cover for the increasing maintenance requirement as the animal grows (this being almost entirely of energy only). The greatest single determinant of variation in the composition of growing pigs is the quantity of food consumed (Fig. 3.12). The linear response of lean tissue growth to increasing food supply is coupled with a minimum level of fat, as is appropriate to normal positive growth. The limited nutrient supply in this phase of the response brings about a constant relationship between fat and lean over a range of growth rates. The minimum fat:lean ratio is thus an important determinant of body composition for much of the life of the growing animal, and appears to be characteristic of sex and breed type (being lower for males and improved strains). Fatty tissue growth above this minimum can only be achieved when food supply is increased such as to exceed the needs for maintenance, maximum potential rate of daily lean tissue growth rate and the minimum balance of fat in normal gain. Feed supply may be said to be unlimiting when the plateau for daily lean tissue growth rate is reached and a concomitant increase is observed in fatty tissue gains as the excess energy is diverted to lipid anabolism. In young pigs of 5–40 kg or so, the physical bounds of appetite usually restrict food supply to the nutritionally limiting range, whilst above 40 kg this is no longer the case. Avoidance of fattening may be achieved by restricting the feed supply in a way that moves it back into the nutritionally limiting phase. (A reduction in fatness below the minimum stated for normal growth would require a severe restriction to bring about fat catabolism.) Pigs with limited appetites may never be able to eat enough to fatten before slaughter weight is attained; while those with lower potentials for lean tissue growth or with higher appetite may need to have their feed intake restricted if fattening is to be avoided (see Table 3.4).

Even whilst feed supply is inadequate to maximize lean tissue growth, there is nevertheless always some deposition of a minimum level of fat commensurate with normal positive daily gains. This minimum may be expressed in terms of a ratio to lean. This minimum fat:lean ratio gives the level of fatness in the pig

Table 3.4 Feeding levels required to prevent undue fattening and the production of unacceptably fat carcasses

	Entire males	Females	Castrated males
Improved strains	*Ad libitum**	*Ad libitum**	Slight restriction
Commercial strains	*Ad libitum**	Slight restriction	Medium restriction
Utility strains	Slight restriction	Medium restriction	Heavy restriction

* Many 'improved' strains of pigs have been bred for leanness by means of selecting for reduced appetite.

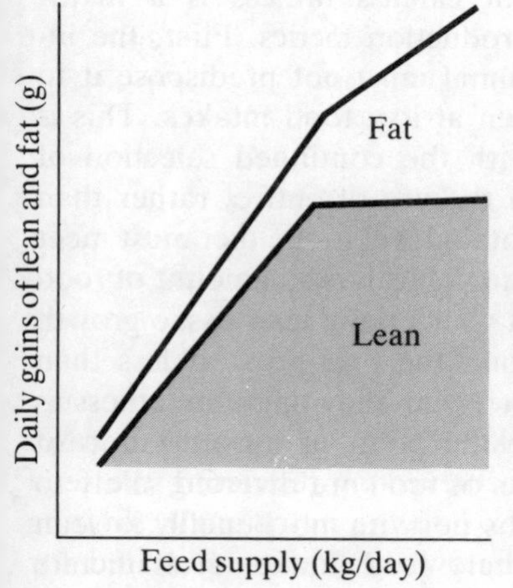

Figure 3.12 Growth responses of lean and fat to increasing supply of feed.

that can only be undercut by creating abnormal conditions of fat catabolism, such as would occur at very slow rates of growth. Should this minimum give backfat depths at slaughter weight that are in excess of the premium grade standard, then only considerable reduction in feed supply and growth rate could place carcasses into the top grade (such might be the case for castrated pigs of low genetic merit). Equally, should the minimum ratio give backfat depths at slaughter weight that are below a minimum fatness standard (as might be the case with entire males of high genetic merit), then adequate fatness can only be achieved by feed intake levels in excess of those that will maximize lean tissue growth rate. The higher the potential for lean tissue growth rate, the more difficult such a feed intake level would be to achieve.

Pigs of high genetic merit may have higher lean tissue growth rate potentials, lower minimum fat ratios, or both. Such animals will be thinner at low feed intakes and more difficult to fatten as feed level increases. Excessive fatness is feasible for all pigs, but only provided that voluntary feed intake is sufficient to put the maximum limit to potential lean tissue growth rate into range. It is evident that, in order to optimize production and achieve grading targets, the feed supply that will maximize lean tissue growth without generating excessive fat is crucial. It is likely that this point is out of appetite range for many young pigs, unless they are managed well and encouraged to eat; equally, for many older pigs a ration may need to be imposed unless the animal is either of high genetic merit or has a low appetite.

Conclusion

The early identification of causes of variation in fatness in pigs has resulted in the breeding of improved types with high maximum daily lean tissue growth rate potentials, low minimum fat ratios in nutrient limited growth and restricted appetites. Specification of nutrient requirements by the factorial method allows exact nutrient supply and avoids the possibility of fatness through faulty diet specification while, where necessary, feed restriction can further control fatness. The minimum amount of fat commensurate with meat quality and consumer acceptability can therefore readily be obtained. The ever narrowing range of fatness that defines the limits of consumer needs is now creating problems for producers, who have to cope with natural biological variation in weight and fatness, whilst simultaneously satisfying targets of diminishing size for both characters. Further reductions in biological variation are most likely to come from improving the nutrition and management of the pig in the early growth phase, where there is currently the greatest gap between achieved performance and biological potential.

The nutritional manipulation of carcass fatness is a major element in optimizing pig meat production tactics. First, the inherent genetic make-up of the animal must not predispose it to unacceptable levels of fatness, even at low food intakes. This is becoming increasingly unlikely with the continued selection of improved strains of pigs and with the use of entire, rather than castrated, males. Secondly, the protein level in the diet must meet the appropriate energy:protein ratio. Thirdly, the amount of food given should relate to the point at which daily lean tissue growth rate is maximized. Below that point the pigs grow at less than optimum speed, but will not be fatter than the minimum fatness as inherently predetermined. Above the point of maximization of lean tissue growth rate, increments of food are diverted solely to the deposition of surplus fatty tissue with no benefit to lean growth. Invariably, limited voluntary feed intake levels dictate that maximum lean tissue growth rates are usually not achieved in commercial pigs of below 40 kg live weight. This situation can be eased by improving both the feed and the feeding method. Once voluntary feed intake no longer limits lean tissue growth rate, as may often be the case above 50 kg live weight, then the producer can restrict feed intake to maintain minimum fatness, or feed more to exceed it. Reduction of feed input will slow down growth rate, reduce fatness, and may improve grade and individual pig value. But the cost benefit of reduced feed inputs requires careful assessment, particularly where throughput and overall feed efficiency is important to production optimization. Where problems

Figure 3.13 *Pigs can now be produced that are very lean. The top bacon rasher (a) has 3 mm of skin and 4 mm of fat, giving a P2 of 7 mm. An alternative strategy is slaughter at a heavier weight (or use a blocky type, or both), which will give a bigger eye muscle (b) but also predisposes to more fat. The undesirable fat can be trimmed back to 10 mm depth by removal of the skin and the first layer of fat (c). This latter pig is likely to have had a P2 fat measurement in excess of 16 mm P2. Both types of product (a) and (c) are readily available in UK shops. Many customers prefer the trimmed rasher with the larger muscle, although the pig giving rasher (a) was by far the most lean.*

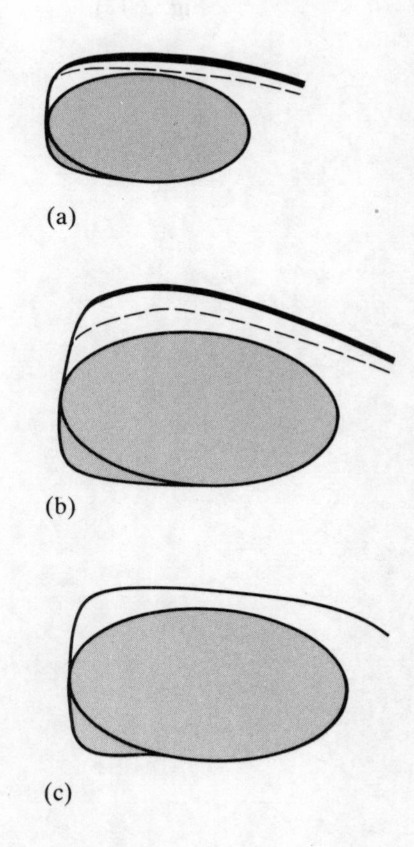

of over-leanness pertain, there is a clear indication for enhancing feed intake to the highest possible level. Pigs of high genetic merit will also carry the potential for high daily lean tissue gains. Again, the consequence is that higher feed levels can be consumed with positive responses to growth rate without forgoing leanness.

Achieving grading standards or grading targets by nutritional manipulation is therefore relatively straightforward. But it should not be assumed that control of the production process is always the most effective route to the provision of customer demands (see Fig. 3.13).

Chapter 4 Energy and protein evaluation of pig feeds

Introduction

Performance responses of growing and breeding pigs to the energy and protein in any given diet is related to the amount of food eaten and to the concentration of digested, metabolized and subsequently used nutrients in the diet.

A view of utilizable energy and protein content in a diet allows calculation of nutritional and monetary worth, and can give a guide to proper rationing and nutrient provision. The expected performance of growing pigs can be predicted from nutritional content, expressed as energy, essential amino acids, total protein, vitamins and minerals. For these purposes, ingredient source of the nutrient is considered unimportant. If the efficiency of use of absorbed energy and protein is known, then growth rate and composition of the growth can be closely predicted. Net utilizations are best calculated through a knowledge of the intermediary metabolism, because direct determination is highly dependent upon the particular animal and the particular environment that prevails at the time of measurement, and therefore liable to be specific rather than general in its view of nutritional content of a feed ingredient. Such specificity will lead to error if the feed ingredient comes to be used for a variety of types of animal in a variety of environments.

For most purposes, an effective mode of operation is to determine the basic level of energy in the diet as the gross energy (GE), the basic level of protein as nitrogen multiplied by 6.25 and the content of amino acids by direct individual chemical analysis. Next, knowledge of the digestibility of the energy and protein fractions is required, and this character may be considered as a general property of the feed which can be applied in a relatively wide variety of situations. Subsequent utilization by the animal of absorbed energy, protein and amino acids is as much a function of the animal in question as the foodstuff, and measurements of metabolizability and net usages are therefore not properly considered as a part of nutrient evaluation; although, of course, they are a part of the final assessment of nutritive worth and are essential to final performance prediction. The best route through the intermediate stages between digestion and incorporation of nutrients into the components of growth is via simulation models. It would be quite impossible to consider the enormity of the experimental design that would be needed to account for all possibilities of perceived interactions in order to determine empirical statements as to the efficiency of utilization of digested nutrients by the animal body for maintenance, lactation, pregnancy and growth. Given a knowledge of absorbed nutrients and of the intermediary metabolism that will pertain, net usages can be readily and accurately calculated by simulation modelling. Examples of such models are described later in Chapter 6.

The objective at this point is to arrive at the means of evaluating pig feeds by the estimation of digestible energy, of digestible protein and of the content of properly balanced amino acids.

Energy

Expression of energy value in food by measurement of digestible energy (DE) takes the simplistic line that the estimate does not concern itself with efficiency of energy use endogenously, but only with the difference between GE ingested and GE excreted from the body in faeces. Thus not only does this measurement not give an accurate picture of energy value as perceived by the animal, but it also makes no correction for previously absorbed energy passing from the body-proper back into the gastro-intestinal tract through the medium of intestinal juices, enzymic secretions, sloughed cells and so on, and subsequently failing to be reabsorbed; and therefore appearing to have failed to be digested.

The measurement does, however, give a view of the level of apparently digested energy in a feed, and does so in a way that is independent of animal effects, and therefore realistic for use as a description of the energy value of a feed in a range of different circumstances and at different times. The DE of diet, in contrast to metabolizable energy (ME) is not confused by the level or quality of dietary protein supply in relation to the ability of the pig to accrete lean tissue (excess dietary nitrogen reducing ME through enhanced urinary excretion rate).

The determination of energy value of feeds by DE measurement (rather than GE, ME or net energy (ME)) has been found particularly useful because it is quite readily made and has tended to yield consistent relationships with achieved animal performance. This has been because pig diets usually comprise a conservative and restricted range of ingredients, so the coefficients relating DE to ME and ME to NE are relatively stable. But, these relationships may weaken if the patency of the intestinal tract changes over time, or a more liberal range of diet ingredients is used. The digestibility and use of complex carbohydrate components will certainly differ when examined in baby and then in adult pigs. Apparent digestibility further merely relates to disappearance, saying nothing of the nature of the end-product disappearing nor of its usefulness. Thus similar DE values would be given to simple sugar energy disappearing from the ileum as glucose and to complex carbohydrate energy leaving the caecum as volatile fatty acids, or even the anus as gaseous methane escapes; which is clearly a simplification, which may result in an important degree of error. A position shift in the usual proportion of dietary components digested in the large intestine rather than the small intestine is likely to diminish the relationship between dietary DE and animal response. Workers at the Rowett Research Institute have proposed the use of a correction factor for fibrous feeds, bulky feeds and otherwise indigestible feeds. This correction is based upon the proportion of energy disappearing from the large, rather than the small, intestine. Thus, while 14 MJ/kg dry matter (DM) of both barley and grass are digested from the gut, the ascribed DE value for grass is 3 MJ less than barley at only 11 MJ, because nearly half of the digested energy is absorbed from the large intestine, whereas the equivalent value for barley is less than one-fifth.

These caveats should not, however, unduly detract from the use of DE, which remains the most effective means of classifying individual foodstuff ingredients and mixed feeds in order to allow predictable pig performance, monetary evaluation of diets and diet ingredients, proper incorporation of ingredients into mixed diets at the correct levels to give a balanced feed, and the ultimate rationing of compounded feed to the pigs in the form of the final total dietary allowance.

At the end of the day, whilst DE is the chosen level for feed description, when it comes to determining energy utilization it is NE that gives the only final and definitive statement of energy balance. NE is dependent upon the nature and activities of both the animal and its environment. The calculation of NE yield from DE supply can be achieved through a series of fixed or semi-fixed conversion factors or, better, by dynamic estimation of nutrient utilization through the medium of a simulation model for prediction of nutrient flow and production response within the body.

Determination of digestible energy by use of live animal studies

The determination of DE is achieved by feeding a measured amount of the given feed material to a pig, crated in such a way that the faeces can be separated from the urine, quantitatively collected and then preserved to avoid energy losses after excretion (Fig. 4.1).

Sample collection may be undertaken with the help of a marker; the concentration of marker giving a means of determining the proportion of the whole represented by the sample. Markers themselves have problems, indigestibility itself being an insufficient quality to ensure exactly proportional excretion. In any event, the need is not great, as crating is no particular problem and total faecal collections for 1 week or so no particular hardship for either pig or scientist. It is essential, of course, that animals are adequately habituated to the material being given (this can range from 3 days to more than 3 weeks) and for the collection to last long enough to avoid the problems of end-effects. At Edinburgh a continuous period of 10 days is usually employed for total faecal collection under acid. (In the case of specific work on fibre or lipid digestibility, which would be disrupted by the acid medium, the material is collected fresh daily and frozen.)

Level of feeding, an important consideration in ruminant balance trials where rate of passage may significantly affect digestibility, does not appear greatly to affect digestibility in growing pigs. However, whilst this may have been exhaustively studied and well proven with young pigs given concentrate diets, the matter is inadequately researched for adult pigs, for pigs given diets high in roughage and for pigs given diets of low DM or based on liquid by-products.

The DE of a diet may be determined by presenting to a crated pig an exact amount of food daily over a given predetermined period and collecting all the faeces excreted during the final 10 days of that period. Normally, the level of feeding will be at least maintenance, but not so much that there is a danger of refusals (this occurs at around about three times maintenace). For a

Figure 4.1 *Metabolism crate for balance studies. Faeces fall into the bucket and urine into the box. Each container holds dilute acid. The food trough reduces spillage, but there is also a tray below. The floor of the cage on which the pig stands is of mesh. The crate is suitable for pigs of 20–100 kg and of male sex. (From Whittemore, C. T. and Elsey, F. W. H. (1976)* Practical Pig Nutrition, *Farming Press, Ipswich.)*

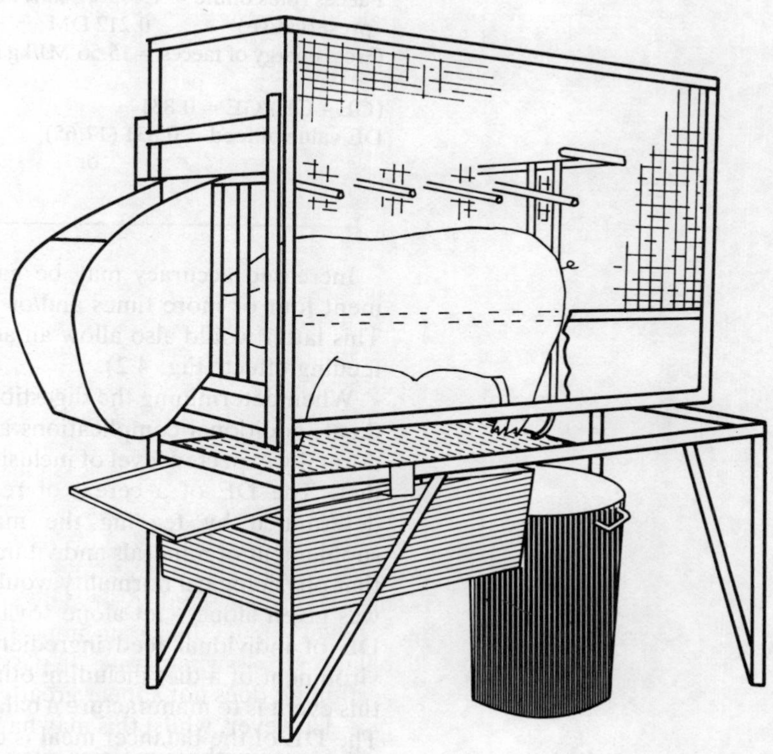

40–50-kg crated pig, 1500 g/day given in two separate meals might be appropriate. GE determinations are carried out on both ingested (I) and faecal (F) material. Where GE_I is the MJ of GE ingested daily and GE_F is the daily faecal excretion, then $(GE_I - GE_F)/GE_I$ gives the digestibility coefficient which, when used with GE concentration of the ingested material, will give the MJ DE per kilogram (fresh or dry) of the diet in question (see Table 4.1).

Table 4.1 Example energy balance for determination of digestible energy (DE)

Intake (kg/day) = 1.500 kg fresh feed at 0.870 dry matter (DM)	= 1.305 kg DM
Gross energy of feed = 17.65 MJ/kg DM; GE	= 23.03 MJ
Faeces (plus dilute preservative) = 1.179 kg slurry at 0.212 DM	= 0.250 kg DM
Gross energy of faeces = 15.56 MJ/kg DM; GE	= 3.889 MJ
(GE – GE)/GE = 0.831	
DE value of feed = 0.831 (17.65)	= 14.67 MJ DE per kg DM
or	= 12.76 MJ DE per kg of fresh feed

Increased accuracy may be gained from replicating the treatment four or more times and/or by feeding three or more levels. This latter would also allow an additional check against a level of feeding effect (Fig. 4.2).

When determining the digestibility of the energy of a diet ingredient, additional complications arise in inverse proportion to the normally expected level of inclusion of the ingredient in a balanced diet. The DE of a cereal of reasonable protein level could be determined by feeding the material alone, or with a small supplement of minerals and vitamins. However, it is inconceivable that physiological normality would pertain if fish meal or rice bran was given alone – let alone soya oil or tallow. Usually, then, the DE of individual feed ingredients is determined within the environment of a diet including other ingredients. The procedure in this event is to manufacture a balancer meal for the test ingredient. The DE of the balancer meal is determined as a first step and the DE of the balancer plus test ingredient as the second step. The difference between the two estimates gives the contribution of the test ingredient. Suppose the GE of the test (T) ingredient is 22.0 MJ/kg DM, and the GE of the basal balancer (B) is 18.0 MJ/kg DM. Firstly, 2 kg DM of B is given, providing 36 MJ of GE per day. Of this, 7.2 MJ GE daily is recovered in the faeces. Next, 0.5 kg of DM of T is given, in addition to the 2 kg of B. This amounts to 47 MJ of total diet GE (36 from B + 11 from T). Of this, 8.3 MJ GE is recovered in the faeces. As 7.2 MJ was recovered from B, then the energy recovered from T was (8.3 – 7.2)

Figure 4.2 *Estimation of digestibility using three levels of intake. A negative effect of intake upon digestibility would be shown by curvilinearity (b). The slope gives the digestibility coefficient (y/x = 0.80). The constant term should be equal or close to zero.*

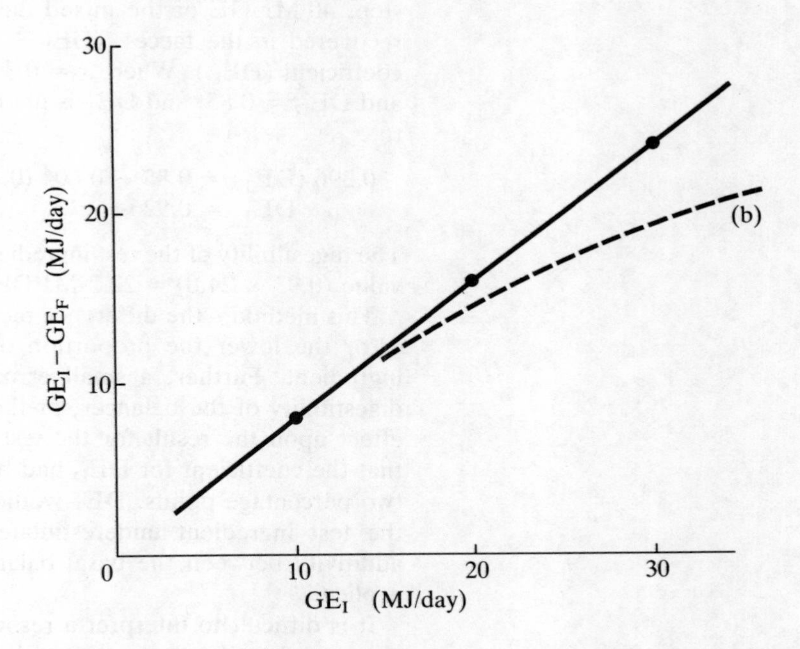

= 1.1 MJ. It can now be concluded that, as 11 MJ of T were given, the digestibility of T is $(GE_I - GE_F)/GE_I$, $(11 - 1.1)/11 = 0.90$. The DE of T is therefore $(0.90 \times 22.0) = 19.8$ MJ/kg DM.

When the test ingredient is added into, rather than on top of, the basal meal, then the calculation must, of course, be weighted by the relative contributions of the basal balancer and the test ingredient to the final diet; bearing in mind that it is not the proportional contribution to the DM that is of concern, but the proportional contribution of the gross energy.

$$x(DE_T) + y(DE_B) = DE_D \qquad \text{[Eqn 4.1]}$$

Suppose the GE of the test (T) ingredient is determined as 24.0 MJ/kg DM and the GE of the basal balancer (B) as 18.0 MJ/kg DM, and the two materials are mixed together in the DM proportions of 0.33 and 0.67. The GE of the mixed diet (D) is

7.92 + 12.06 = 19.98. Of this, 0.396 is from the test ingredient and 0.604 from the basal balancer. On the first step, 36 MJ GE of B is given daily and 7.2 MJ GE recovered in the faeces. $(GE_I - GE_F)/GE_I = 0.80$ digestibility coefficient (DE_B). On the second step, 40 MJ GE of the mixed diet is given daily and 6 MJ GE recovered in the faeces. $(GE_I - GE_F)/GE_I = 0.85$ digestibility coefficient (DE_D). Where $x = 0.396$ and $y = 0.604$; $DE_B = 0.80$ and $DE_D = 0.85$; and DE_T is unknown; then equation 4.1 solves to:

$$0.396\ (DE_T) = 0.85 - 0.604\ (0.80)$$
$$DE_T = 0.926$$

The digestibility of the test ingredient is therefore 0.93 and the DE value (0.93 × 24.0) = 22.2 MJ DE per kilogram DM.

This method – the difference method – is clearly more prone to error the lower the proportion of GE contributed by the test ingredient. Further, a small error in the determination of the digestibility of the balancer, or the mixed diet, can have a large effect upon the result for the test ingredient. Say, for example, that the coefficient for DE_D had been underestimated by merely two percentage points. DE_T would be 0.88 and the DE value of the test ingredient underestimated by fully 5 per cent. Last, additivity between the basal balancer and the test ingredient is explicit.

It is difficult to interpret a response when the inclusion of the test ingredient has been achieved simultaneously with, and at the expense of, a withdrawal of the equivalent amount of basal balancer. One is studying two events and not one. This becomes especially important if the type of energy in the balancer (say, starch) is suspected of being physiologically different from the type of energy in the test ingredient (say, protein), or if a chemical component of the energy fraction is being studied, such as would be the case for an examination of the digestibility of neutral-detergent fibre (NDF).

These problems can be greatly diminished by the procedure used now for many years at Edinburgh. To a single level of basal diet are added incremental levels of test ingredient. In one experiment, to 1 kg of basal balancer (B) given daily was added 0, 0.2, 0.4 and 0.6 kg of test ingredient (T_1), and again 0, 0.2, 0.4 and 0.6 kg of a second test ingredient (T_2). Three animals were used at each level. The type of plots obtained were as shown in Fig. 4.3.

Figure 4.3 *Determination of digestible energy (DE) of test ingredients T by means of regression.*
For T1: Y = *13.1 + 16.5*X(●)
For T2: Y = *12.9 + 11.8*X(■)
Where 1 kg of basal balancer (B) given, then $DE_B = 13.0$ MJ/kg, $DE_{T_1} = 16.5$ MJ/kg and $DE_{T_2} = 11.8$ MJ/kg.

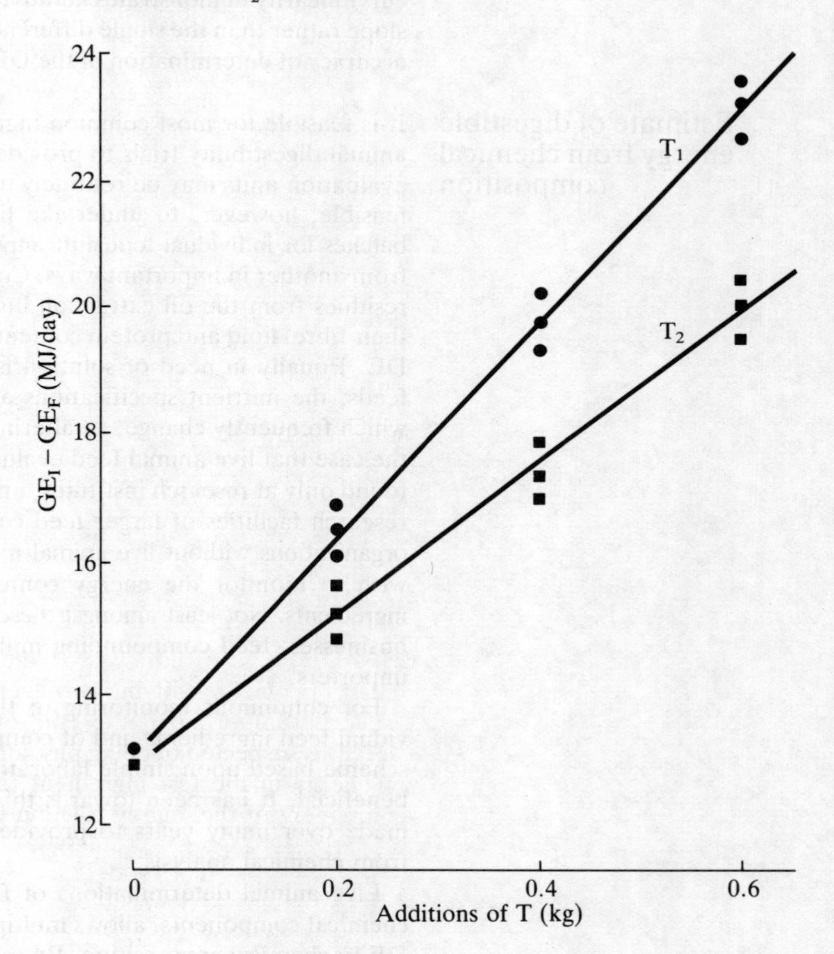

The slope of the line indicates the DE value of the added test ingredient (T), and the constant the DE value of the basal diet (B). Similar values for B in the cases of both experimental procedures for T_1 and T_2 gives reassurance as to the correctness of that value in different dietary combinations, whilst the lack of curvilinearity demonstrates additivity. The use of the regression slope rather than the single difference estimate gives much greater accuracy of determination of the DE value of T.

Estimate of digestible energy from chemical composition

It is feasible for most common ingredients to be run through live animal digestibility trials to provide updated DE values; and feed evaluation units may be routinely used for such purposes. It is not feasible, however, to undertake live animal trials on individual batches for individual foodstuff importations, which may differ one from another in important ways. Common cereals and protein-rich residues from the oil extraction business often display a range in their fibre, lipid and protein contents, with consequent effect upon DE. Equally in need of solution is the problem of compounded feeds, the nutrient specifications and ingredient compositions of which frequently change, so altering the final DE value. It is also the case that live animal feed evaluation units are invariably to be found only at research institutes, university departments and some research facilities of larger feed compounders. Very many other organizations without live animal metabolism facilities nevertheless wish to monitor the energy content of pig feeds and pig feed ingredients. Not least amongst these are pig producers, agricultural businesses, feed compounding mills, and commodity buyers and importers.

For continuous monitoring of the nutritional content of individual feed ingredients and of complete compounded pig feeds, a scheme based upon simple laboratory chemical analysis would be beneficial. It has been towards this end that attempts have been made over many years to provide prediction equations for DE from chemical analysis.

Live animal determinations of DE, together with analysis for chemical components, allows multiple regression analysis to relate DE to chemical composition. Regression is a particularly effective means of stating in as closely exact a mathematical way as possible relationships found in any particular set of data. Regression itself does not, however, imply effective prediction. If regression is to be used for forward prediction, rather than historic description, then

Figure 4.4 Choice of regression line for accurate prediction of DE of edible grade oil:
(a) slope = 0.0100; k = *20.0*
(b) slope = 0.0215; k = *17.5*

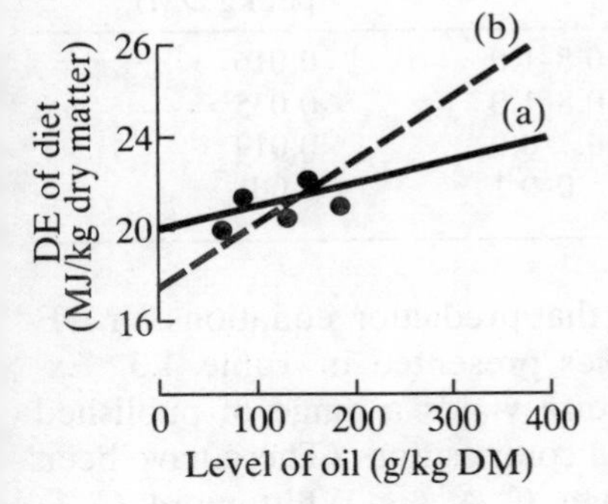

accuracy is greatly helped if the variables in the equation are themselves the causal forces of the variation in the parameter to be predicted. Further, the coefficients and constants should have biological logic and relevance. Should these requirements not be met, a regression equation is in danger of giving erroneous predictions, regardless of the numerical accuracy with which it describes the data set from which it was constructed. Errors become the more likely when the prediction is required to relate over a wider range than the original data, or to differing combinations of ingredients.

Regression of the data set in Fig. 4.4 may, for example, give a line of best fit (a) with a constant term of around 20 and a slope of 0.010, indicating each additional gram of oil to increase DE by 0.010 MJ. Such a line states: (i) oil has a DE value of about 30 and (ii) diets with no oil will have a DE of about 20. These propositions require to be checked against the common sense that (i) the DE of a diet without oil should approximate to the expected DE of the remaining components (fibre, starch and protein), and (ii) that oil is highly digestible to the pig and will have a DE value not dissimilar from the GE. Such ideas as these latter would accord much more closely with a line (b) of slope 0.0215 and constant term 17.5

The major causal forces of variation in DE concentration in the DM of foodstuffs may be put forward as: a negative contribution from the presence of fibre; a positive contribution from the presence of oil; no contribution from the presence of carbohydrate (starch contributing primarily to the average DE of the diet, rather than to deviations from that average); and a small positive effect of protein. Active variables in useful regression equations should bear some relationship to the above suppositions.

Expectations for the form of an effective regression might be something as follows:

1. A constant term close to zero if either gross energy or carbohydrate is included in the equation.
2. A constant term close to the average DE of the diet if carbohydrate or gross energy are not included in the equation.
3. Fat and fibre should have large effects within the equation, with protein (and ash) contributing less.
4. Coefficients should bear some relationship to the logic set out in Table 4.2.

Table 4.2 Expected contributions of chemical components to the concentration of diet DE

Chemical component	Approximate GE (MJ/kg)	Assumed digestibility	Contribution made by 1 g (MJ DE per kg DM)
Starch	17.5	0.8–1.0	0.016
Oil	39	0.8–1.0	0.035
Protein	24	0.7–0.9	0.019
Fibre	17.5	0–0.1	0.001

It may be postulated therefore that prediction equations for DE may approximate to the examples presented in Table 4.3. Examination of the research literature yields a range of published equations relating DE to chemical composition. (These have been reviewed and referenced by Morgan, C. A. and Whittemore, C. T. (1982) *Animal Feed Science and Technology* **7**: 387.) Some examples of the genre are given in Table 4.4. The first equation has a negative multiplier for crude fibre that is higher than might be expected. This suggests either that crude fibre does not measure all the indigestible fibrous components negatively active in the diet, or that the presence of crude fibre disrupts the digestibility of other diet components. The constant term in this equation is also a little low, implying that a diet with no ether extract and no crude fibre (i.e. made of carbohydrate and protein) has a DE value of 16.0 MJ/kg DM. The second equation would please phlogistonists as it appears that, upon digestion, more GE is yielded than was ingested. This is then corrected by a negative constant term. The third equation was, as it happens, a remarkably accurate description of the data set in question, but has unexpectedly high multipliers on crude protein, ether extract and nitrogen-free extractives (NFE). Fibre is absent from the equation, but is included by implication via NFE (which is, of course, a difference value). These issues are resolved in statistical terms by a weighty negative constant.

Table 4.3 Postulated forms for equations to predict DE from chemical components

1. DE (MJ/kg DM) = 0.016 starch (g/kg) + 0.035 oil (g/kg) + 0.019 protein (g/kg) + 0.001 Fibre (g/kg).

2. DE (MJ/kg DM) = 17.0* + 0.018 oil† (g/kg) + 0.002 protein (g/kg) − 0.016 fibre‡ (g/kg) − 0.001 starch (g/kg)

* 70% starch, 20% protein, 5% oil, 5% fibre. This term may also be expressed by using GE (MJ/kg) multiplied by the expected overall digestibility coefficient for GE, e.g. 0.84 GE

† Removal of 1 g of average DE and substitution of 1 g of oil DE: − 0.017 + 0.035 = 0.018

‡ Removal of 1 g of average DE and substitution of 1 g of fibre DE: − 0.017 + 0.001 = − 0.016

Table 4.4 Three example prediction equations drawn from the literature (all elements per kg DM)

1. DE (MJ/kg DM) = 16.0 − 0.045 CF(g) + 0.025 EE(g)
2. DE (MJ/kg DM) = − 4.4 + 1.10 GE (MJ) − 0.024 CF(g)
3. DE (MJ/kg DM) = − 21.2 + 0.048 CP(g) + 0.047 EE (g) + 0.038 NFE(g)

CF = crude fibre; EE = ether extractives; CP = crude protein; NFE = nitrogen-free extractives

The Edinburgh experiments

With these matters in mind, a series of experiments was undertaken at Edinburgh between 1980 and 1985. These are still continuing, with a particular examination of the role of fibre in monogastric nutrition.

Acting on the proposition that fibre was the most important component reducing diet digestibility, the relationship between analysed chemical fibre content and digestibility in the live animal was examined using three different fibre types. These were those found in oat feed, rice bran and beet pulp. Oat feed contains about 250 g crude fibre per kilogram DM, rice bran about 160 g and sugar beet pulp about 190 g. The basal balancer used was wheat/soya and, to 1 kg/day of this, graded levels of the test materials were added. By the regression method of plotting DE intake ($GE_I - GE_F$) against level of addition of feed, the following equations were derived:

$$GE_I - GE_F \text{ (oat feed)} = 13.53 + 8.56\,X$$
$$GE_I - GE_F \text{ (rice bran)} = 13.76 + 10.25\,X$$
$$GE_I - GE_F \text{ (beet pulp)} = 13.86 + 11.16\,X$$

The regressions suggest that the DE of the basal balancer is relatively unaffected by the source of added fibre, but that the DE value for the three materials (MJ DE per kilogram, as given) were 8.56, 10.25 and 11.16 respectively.

Crude fibre is a term used to describe an analytical procedure and it has little meaning in terms of structural plant elements. The Van Soest procedures attempt to differentiate between various fibre fractions in a sequential analysis. Lignin is the most complex fibre material, followed by cellulose and, hemicellulose. There is agreement that lignin is indigestible to the small intestinal digestive enzymes of the pig and also indigestible to all other enzymes, even those of bacterial origin found in the large intestine and caecum. It may be supposed that cellulose is indigestible in the small intestine, but potentially digested posterior to the ileum. Argument rages as to the potential and site for the digestibility of hemicellulose. But the weight of current opinion is that hemicellulose may, in fact, be rather less digestible than cellulose, being resistant to degradation in the large intestine and caecum. In the analytical series, lignin is estimated as acid-detergent lignin or alternatively as Christian lignin; acid-detergent fibre (ADF) estimates lignin plus cellulose and neutral detergent fibre (NDF) estimates all three (lignin, cellulose and hemicellulose).

Fibrous components of four example feeds are shown in Table 4.5. Straw is rich in cellulose, oat feed and rice bran have similar structural carbohydrate compositions, while sugar beet pulp has relatively more cellulose and relatively less lignin. The digestibility of the feed material will be consequent upon the absolute amount of total fibrous components, and the balance of hemicellulose, cellulose and lignin within the fibre. Thus the relatively higher total fibre content of sugar beet pulp as compared with rice bran appears to have been more than offset by sugar beet pulp having a much lower proportion of lignin within the fibre. The digestibility of oat feed as compared with rice bran reflects the greater absolute quantities of structural carbohydrate components in oat feed as compared with rice bran.

Table 4.5 Fibrous components of some feeds (g/kg DM)

	Straw	Oat feed	Rice bran	Sugar beet pulp
Chemical composition				
Crude fibre	478	252	155	192
Neutral-detergent fibre	775	579	353	349
Acid-detergent fibre	544	309	188	241
Christian lignin	152	136	76	22
Structural carbohydrate composition				
Hemicellulose	231(0.30)*	270(0.47)	165(0.47)	108(0.31)
Cellulose	392(0.51)	173(0.30)	112(0.32)	219(0.63)
Lignin	152(0.20)	136(0.23)	76(0.22)	22(0.06)

*Values in parentheses give the relative proportions.

When the various ways of analysing for fibrous components were examined with a view to their efficacy in predicting DE, it was evident from both the logical and the statistical point of view that NDF was much more effective than the others, whilst crude fibre (CF) was especially poor. This might be considered as a predictable outcome, as CF is an empirical analysis not based on any particular fibre source (see Table 4.5), whereas NDF includes all three fibre fractions considered to be more or less indigestible in the pig.

The statistically determined best-fit prediction equation based on the combined data for all three fibrous sources was:

$$\mathrm{DE\ (MJ/kg\ DM)} = 17.4 - 0.016\,\mathrm{NDF}$$

which accords remarkably well with the propositions laid out in Table 4.3.

Acting on the proposition that oil was the most important component increasing diet digestibility, the relationship between analysed oil content and digestibility was examined using three different sources of human-edible grade oil: tallow, palm oil and soya oil, added at progressively increasing levels on to a basal balance of 1 kg of a barley/soya diet:

$$GE_I - GE_F\ \text{(tallow)} = 12.90 + 38.82\ X$$
$$GE_I - GE_F\ \text{(palm oil)} = 12.60 + 37.54\ X$$
$$GE_I - GE_F\ \text{(soya oil)} = 12.92 + 40.83\ X$$

In the case of the added oils, the test materials were pure, whereas with fibrous feeds the test materials were high in fibre, but not solely comprising fibre. As the GE of oil is about 39.6 MJ/kg, the regressions indicate very high digestibilities indeed. The possible expectation that the more highly hydrogenated tallow was not as digestible as the unsaturated palm and soya oils was not realized.

The best fit regression over all oils produced a positive coefficient of +0.025 for ether extractives (EE) (g/kg), which is higher than the 0.018 proposed in Table 4.3, but not too far distant from the 0.023 that logic would have proposed for a very high-grade oil, which might be completely digestible.

These results gave the Edinburgh workers sufficient confidence to set up a full-scale experiment involving the use of 36 compounded diets covering most, if not all, of the ingredients to be found in pig feeds (33 in total), and giving a composition range of 23–123 g crude fibre per kilogram, 149–258 g crude protein per kilogram, and 21–116 g EE (oil) per kg. These diets were each given to four growing pigs and the complete data set used for multiple regression analysis to derive equations for the prediction of DE from the chemical analysis of compounded diets.

A very large number of equation forms are possible and of the 62 of particular interest the following have been drawn from the source report (Morgan, C. A. *et al*. (1984). The energy value of compound food for pigs. Edinburgh School of Agriculture and AFRC Unit of Statistics).

1. **Using declared analysis:**

$$DE\ (MJ/kg\ DM) = 18.3 - 0.037\ CF - 0.019\ ash + 0.011\ oil.$$

This is not particularly accurate (residual standard deviation = 0.65), neither are the coefficients particularly logical. The high multiplier for CF may indicate either that CF estimates only a part of the indigestible diet constituents or that the presence of CF disrupts digestion of other non-fibrous diet components. If either of these possibilities pertain, the naive use of an elevated multiplier is unlikely to be sufficient means of ensuring the accuracy of a prediction equation. In any event, it was perfectly clear from the whole exercise that, of all the analyses for fibre, for the purpose of DE prediction, NDF was considerably superior to any other measure, and CF was particularly inferior.

2. **Including gross energy:**

$$DE\ (MJ/kg\ DM) = 3.77 - 0.019\ NDF + 0.758\ GE.$$

This equation is rather accurate, with a residual standard deviation of only 0.38. The coefficient for GE reflects the digestibility coefficient itself, although it is less than the mean digestibility, due to the presence of a positive constant term of 3.77 MJ. The coefficient for NDF is within expectation. Unfortunately, many analytical laboratories are not well enough equipped to be sure of an accurate measurement of GE.

3. Using NDF and oil together with a constant term:

DE (MJ/kg DM) = 17.0 − 0.018 NDF + 0.016 oil.

This equation is quite accurate, having a residual standard deviation of 0.44, and has the benefit of restricting itself to the two major causal forces of digestibility: the negative force of fibre and the positive force of oil. It conforms with the propositions put forward in Table 4.3. Further, the coefficient for NDF accords with the expectations derived from the first Edinburgh fibre experiment. The coefficient for oil is lower than that in the first Edinburgh oil experiment, where the three human-grade oils were used. In that previous case the coefficient was 0.025, as compared with 0.016 here. The value of 0.016 is, however, more in line with expectations for feed-grade added oils and oils naturally occurring in feed ingredients; these latter being the effective contributors to oil in the commercial-type compounded diets used in the 33-diet matrix experiment here. The coefficient of 0.016 for oil indicates an achieved DE of dietary fats of around 33 MJ/kg. For simplicity and logic, this equation has much to commend it.

4. Using NDF, oil, ash, protein and a constant term:

DE (MJ/kg DM) = 17.5 − 0.015 NDF + 0.016 oil + 0.008 crude protein − 0.033 ash.

This equation is the most accurate of all, having a residual standard deviation of only 0.32. The inclusion of ash and crude protein (CP) helps to make the equation more robust over a wide range of circumstances.

Protein

Although differences exist in the apparent digestibility of CP (nitrogen × 6.25) of pig feed ingredients, the concept of digestible crude protein (DCP) has not been readily taken up by feed formulators. This is probably because variation in digestibility is (wrongly) not considered as important for protein as for energy. The digestibility of most proteins of animal origin (milk, fish and meat) is about 0.90 and of most proteins of vegetable origin (cereals, legumes, roots and young leaves) about 0.80. Not withstanding the reluctance of some nutritionists to use DCP as the operational standard, protein digestibility has an important influence on feed value, and is profoundly affected by both extrinsic and intrinsic factors.

Usually the digestibility of protein determined as $(CP_I - CP_F)/CP_I$ is some 10 per cent higher than the digestibility determined at the point of exit of digesta from the terminal ileum. Much of the reason for this is the disappearance of ammonia and amines from the large intestine. These nitrogenous materials count toward CP, but not toward protein proper, going directly to urinary excretion. Where the measured DCP value may therefore be in the region of $0.85 \times$ CP, the effective DCP value may well be nearer $0.75 \times$ CP.

Some factors influencing digestibility

The basic amino acids that make up protein are highly digestible. Nevertheless, in the small intestine there is more protein passing down the gut than was eaten. This is because the gut contents are inundated with proteinaceous secretions (mostly enzyme-containing fluids), and also the cells lining the intestine have a high turnover rate and are lost in quantity into the gut lumen. Fortunately, reabsorption of these secretions is as highly efficient as the primary absorption of the amino acid products of enzyme degradation of diet proteins.

Heat damage. Digestibility falls dramatically if the protein is heat damaged: such damage may occur when either animal or vegetable proteins are overcooked as a part of the production process. The consequences of such damage are no part of response prediction, nor of quantitative nutritional evaluation of feeds. Suffice to say that damaged materials – once identified – should be rejected as sources for amino acids for pigs. If used, on the grounds of economic expediency, the effects upon growth and reproduction would be difficult to elucidate other than by empirical trial.

Heat damage changes the protein structure, which reduces its digestibility. The greater the degree of heating, the higher the loss in digestibility. To some extent, all cooked proteins are less digestible than raw ones (about 5% less). But in cases of over-cooking the matter becomes serious and digestibility may be reduced to 50 per cent or less as, for example, in overheated soya bean meal, fish meal, and meat and blood meal. Specific amino acids may be bound on heating. The classic case is the binding of lysine to sugar compounds: this reduces the digestibility and utilizability of lysine. 'Available lysine' is particularly important: (a) because it is a general indicator of heat damage; and (b) in consequence of cereals being lysine deficient, lysine is the most valuable amino acid in most supplementary protein sources. If the lysine is unavailable, then all the other amino acids in the protein, even though utiliz*able*, cannot actually be utiliz*ed*.

Fur, feather and hide. The protein may be bound in the feed in a form resistant to enzyme attack as, for example, the proteins in leather and feather meal.

Abrasion. Certain foodstuffs may increase the rate of cell wall loss from the gut lining and/or prevent efficient reabsorption of this protein of body origin. This will increase the protein in faeces and decrease perceived digestibility. Straw has been implicated in such activity, and also other highly fibrous and abrasive feeds. The evidence is, however, not straightforward.

Rate of passage. Digestive enzymes need time to work. Digestibility can therefore be reduced by anything that increases the rate of passage of digesta through the intestine. High feeding levels of liquid diets of low DM based on milk by-products or liquid wastes may produce just such an effect.

Protection. Enzymes also work most efficiently if they have a large surface area to act upon. Digestibility will be reduced if feed material is only coarsely ground and in large fragments, or if the diet contains large masses of material, which may surround the protein moieties and prevent enzyme penetration. Proteins may be found mixed up with cell wall and other poorly digestible carbohydrate fractions. Until these resistant carbohydrates have been broken down, the proteins cannot be released and will remain undigested. Similarly, protein constituents of cells will only be available for enzyme attack and absorption after such time as the cell wall has been digested away, or ruptured mechanically (by milling, grinding, biting or chewing) prior to reaching the small intestine. Given that a diet contains no toxic protease activity, is not heat damaged and is pulverized to a reasonable extent, protein digestibility appears to be primarily related to the amount of structural carbohydrates present (Table 4.6).

Table 4.6 Digestibility of protein in some foodstuffs

Foodstuff	Digestibility (%)
Fish meal	90–95
Soya bean meal	85–90
Maize meal	80–85
Wheat meal	80–85
Barley meal	75–80
Wheat offals	40–70

These values relate to the apparent digestibility (I−F)/I. Ileal digestibilities are about 10 percentage units lower. In terms of usable protein absorbed, the ileal value is that that should pertain

Anti-nutritional factors. Most, if not all, plant proteins are associated with anti-nutritional factors. These are present to very different extents in different plant species and even in different varieties within species. Thus cereal protein is low and pea protein is quite low in toxic factors, whilst soya bean is rich in protease inhibitor activity, and potato rich in inhibitors and, if green, also in alkaloids. Winter sown rape seeds can be goitrogenic, and contain generous levels of glucosinolates and erucic acid, while in some spring-sown varieties these poisons are at much lower levels. Goitrogens, tannins, saponins, gossypols and alkaloids are heat stable, but the lectins, protease inhibitors and other poisonous amino acids are destroyed by heat. Heat treatment remains the most effective way of counteracting heat-labile anti-nutritive factors. Such processing requires careful control in order to achieve adequate cooking to detoxify, but to avoid overcooking and heat damage; this balance is crucial to the effective production of soya bean meal following the extraction of soya oil. Heat-stable poisons can only be extracted by complex exchange technology or, more commonly to date, by breeding improved strains of plants not carrying the offensive material. The most notable recent case of breeding toxins out of plants has been that of the spring-sown Canadian rapes.

Protease, or specifically trypsin, inhibitors act as anti-enzyme factors, reducing protein digestibility in the gut. In some foodstuffs, levels of typsin inhibitor activity are extremely high, particularly in the bean families (field beans, soya beans, etc.) and potatoes. Uncooked soya and potato can have protein digestibilities lower than 30 per cent. These protease inhibitors act not just specifically in relation to the feeds that contain them, but generally. Thus *all* the protein in a diet containing raw soya will have reduced digestibility.

It is because protease inhibitors are themselves made of protein that their activity is destroyable by heat treatment. Partial cooking causes partial destruction only, and it is this gradation of treatment effect that makes quality control so important for soya beans (and for potato) intended for pig diets. One way of counteracting a low level of inhibitor activity is to give luxury amounts of the protein and it has been accepted for many years that diets containing field beans should be compounded to a more generous protein content.

Given that the proteins are not heat damaged and that anti-nutritive factors are either absent or at a trivially low level, the statement of protein value in pig feeds normally rests wholly upon the profile of essential amino acids.

Biological value – utilization of digested protein

Of all the amino acids commonly found in pig diets (around 22), about nine are essential. The essential amino acids cannot be manufactured within the body, whereas the non-essential are interconvertible. Because they cannot be synthesized by the animal, the essential amino acids must be supplied in the diet. Their importance depends on the priority of their need by the pig in relation to their abundance in diets. Thus, in comparison with requirements, valine and isoleucine are relatively abundant in foodstuffs. Lysine, histidene and methionine, on the other hand, are required in larger quantities and are less abundant. Tryptophan can be lacking in maize diets.

The significant points about the essential amino acids are:

- The value of the diet protein, called the biological value, depends on the relationship between the balance of amino acids required (the balance of amino acids in what the pig sees as 'ideal protein') and the balance in the diet. Diet quality, as measured by biological value, is therefore not absolute, but *comparative* to the 'ideal protein' spectrum.

- The total utilizable (or available) protein (Fig. 4.5) depends on the most (or first) limiting essential amino acid in the diet. This controls biological value.
- The amino acid spectra of proteins of individual foodstuffs (their particular bilogical values) are largely irrelevant to the issue: it is the spectrum of the *consumed* dietary mix that is operative. The dietary bilogical value is NOT the average of the biological values of its ingredients, but the consequence of the complimentarity of the amino acid spectra of the ingredients.

Figure 4.5 *Utilization of dietary protein. Efficiency of use of dietary protein is controlled by:*
(a) digestibility;
(b) utilizability (proportion of total digested protein that is of ideal amino acid balance);
(c) maintenance (protein mass and protein turnover);
(d) level of oversupply.

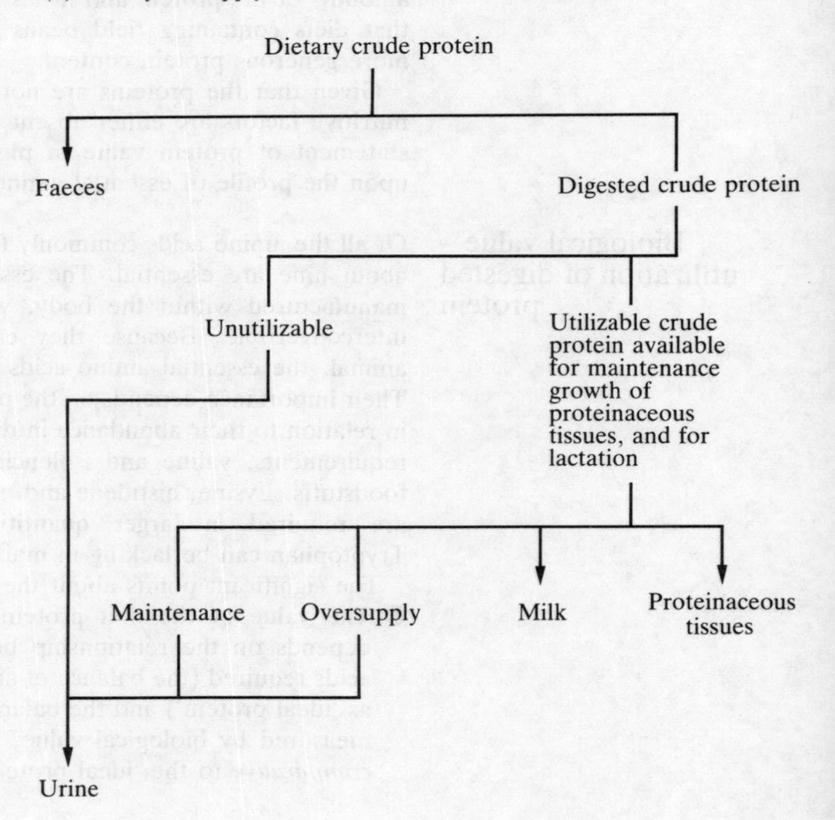

Calculation of biological value first requires the definition of the amino acid balance in 'ideal protein' for pigs. Much painstaking work has gone into the determination of amino acid requirements for pigs: in many American centres and in Poland, particularly at Jouey-en-Josas in France, and also the University of Nottingham, the National Institute for Research in Dairying (now the AFRC Institute of Animal Production and Grassland Research) and the Rowett Research Institute in the UK. In classical nutrition work it had been proposed that egg protein or milk protein represented an ideal amino acid balance. There is perhaps more logic in asking what the amino acids are going towards: namely, mostly meat protein manufacture, with some for maintenance in growing pigs; maintenance in pregnant sows; and milk synthesis in lactating sows. Then, on the assumption that there is little difference in the efficiency of utilization of individual absorbed amino acids, ideal protein should look rather like pig meat protein. Sow milk protein should give some further guide as to the dietary ideal. Fortunately, there is a great similarity between the balance of amino acids in pig meat, sow's milk and the estimates for amino acid requirements determined at research institutes. Perhaps, because there is some inefficiency in collecting up sulphur amino acids during protein turnover (which is to do with maintenance demands), methionine and cystine levels in ideal protein could be higher than in pig protein. The essential amino acid composition of a suggested 'ideal protein' is given in Table 4.7. The total of all the essential amino acids in Table 4.7 seems to be just less than half of the total protein. The other half is the ideal dietary requirement for non-essential amino acids.

Table 4.7 Ideal protein for growing pigs

Amino acid	g amino acid per kg ideal protein
Histidine	25
Isoleucine	40
Leucine	80
Lysine	70
Methionine and cystine	40
Tyrosine and phenylalanine	70
Threonine	40
Tryptophan	10
Valine	50
Total essential amino acids	425
Total non-essential amino acids	575

Having provided the criterion for ideal protein, the amino acid spectrum of the protein in a mixed diet can now be compared with it and the biological value of the diet derived (Table 4.8). The effective biological value of the protein in the diet is that proportion of dietary protein judged utilizable on the strength of the most limiting amino acid. In this case lysine is most limiting and the maximum efficiency of use of this diet protein for growing pigs is 0.64 or 64 per cent. For the diet in Table 4.8, Fig. 4.5 indicates that 0.64 of the protein will be utilizable (although not necessarily all utilized) for maintenance and for protein tissue synthesis, while 0.36 will be deaminated and excreted via the urine. This method for determining biological value takes no account of imbalance amongst amino acids consequent upon oversupply. There is reason to believe that, in the case of some amino acids, oversupply may be a detrimental factor with a negative influence upon biological value. Scoring negatively for imbalance due to oversupply would have the practically realistic effect of reducing biological value in circumstances where the more simple calculation appears to come to an unexpectedly high score.

Table 4.8 Derivation of the biological value of a mixed diet

	g amino acid per kg diet protein (A)	g amino acid per kg ideal protein (B)	A/B
Histidine	20	25	0.80
Isoleucine	40	40	1.00
Leucine	70	80	0.88
Lysine	45	70	0.64
Methionine and cystine	30	40	0.75
Tyrosine and phenylalanine	65	70	0.93
Threonine	30	40	0.75
Tryptophan	10	10	1.00
Valine	40	50	0.80

The biological value is taken as the lowest score in the column A/B

Addition of lysine to the diet examined at Table 4.8 could, theoretically, raise the lysine concentration in the protein to, say, 55 g/kg, in which case threonine, and methionine plus cystine would become the limiting amino acids, and the biological value would become 0.75 (75%). There is, however, still some doubt as to the efficacy of large additions of synthetic lysine to boost biological value. It seems that the efficiency of its use may be lower than natural lysine if the total artificial lysine contributes more than about 0.15 percentage units of total food lysine. Synthetic lysine may be more rapidly absorbed, and arrive at synthesis sites before the other balancing amino acids. It is thus excreted as excess, after which the lysine-deficient amino acid mix comes along too late to use it.

The proportion of ideal protein in cereal protein is usually about 0.5; so for pig diets, protein supplements are added to raise the value of the mixed diets. Protein sources rich in lysine are particularly good at complementing cereal proteins, as shown in Table 4.9. But, although lysine is often the first limiting amino acid in diets and gives the lowest number, this is not invariably the case.

*Table 4.9 Some protein sources used in pig diets grouped according to their ability to impart high biological value to diets in which they are included**

Class 1 (good) protein sources	Class 2 (mediocre) protein sources	Class 3 (poor protein sources
Fish meals	Meat, and meat and bone meals	Cereals
Milk products	Rape seed	Cereal by-products
Single-cell proteins	Groundnut	
Soya bean	Field beans	

Type of diet	Approximate biological value†
Class 3 sources alone	0.45–0.55
Class 3 + class 2	0.50–0.60
Class 3 + class 1 + class 2	0.55–0.65
Class 3 + class 1	0.65–0.75

* This table refers to the quality of the protein in feeds. This is independent of the level of protein, although many feeds high in protein are in the class 1 category and many feeds low in protein in class 2 or 3.

† The higher the inclusion level of any protein source, the greater its effect on dietary biological value.

In sow diets, dietary protein values usually vary between 0.60 and 0.75. In growing pig diets, the range is normally between 0.65

and 0.75. In baby pig diets, the range is normally between 0.70 and 0.80. Even when the calculated number for diet protein value is in excess of 0.85, it is often best to take 0.85 as an assumed pragmatic upper limit for commercial diets.

As a shorthand method of calculation, assuming lysine to be first limiting, Table 4.10 applies. The calculation is, of course, the same as is done in Table 4.8 for each amino acid.

*Table 4.10 Calculation of biological value, from declared lysine level and declared protein level**

CP in diet (g/kg)	170	170	180	200
Lysine in diet (g/kg)	8.3	6.0	7.5	11.2
Lysine in diet protein (g/kg)†	0.049	0.035	0.042	0.056
Calculated biological value (lysine in diet protein ÷ 0.07)‡	0.70	0.50	0.60	0.80

* This calculation applies only where lysine is indeed the amino acid controlling biological value of the diet in question. This is not *invariably* the case.
† Lysine ÷ CP
‡ The calculation is summarized as: (lysine in diet (g/kg)/CP (g/kg))/0.07

Deamination of a proportion of the absorbed diet protein is a beneficial attribute. Indeed, it is one of the functions of pigs to convert lower quality plant proteins into higher quality pig proteins. To do this, the unbalanced amino acids must be excreted. To feed 'ideal protein' in a diet would be grossly uneconomic and it is appropriate to offer more of a lower quality. As the pig requires its protein in terms of daily mass, not diet proportion, one diet protein with half the biological value of another can provide for the same requirement if given in twice the amount. Energy considerations aside, it only requires the second protein to be less than half the price of the first for the poorer quality to be the better buy.

If there is differential absorption of amino acids, then ideal protein expressed at metabolic level will be different from that expressed at diet level. There is some evidence to this effect; thus, the Agricultural Research Council (1981) in its book *Nutrient Requirements of Pigs* (CAB, Slough) states: 'Measurements of amino acid digestibilities are surrounded by considerable uncertainties which should be resolved before a system of feeding based on absorbed amino acids can be advocated'. However, there is insufficient evidence yet available to reject the following simple practical propositions: each individual essential amino acid is approximately equally well digested as any other; all essential amino acids are approximately equally digested as the total diet protein; efficiency of use post-absorption is similar for all essential amino acids; and the ideal balance of essential amino acids is approximately similar for maintenance and for deposition in tissues or milk.

It has been implicit thus far that ideal protein, once absorbed into the bloodstream, will be used with complete efficiency for purposes of protein maintenance, protein growth and milk protein synthesis. Figure 4.6(a) is not reassuring on this score. Many research groups have shown that with increasing supply of ideal protein, response falls away and efficiency of utilization is reduced. This, of course, blows a considerable hole in the notion that a generally applicable economic valuation of feed protein may be achieved by the calculation of ideality. Figure 4.6 requires the use of a diminishing response function based upon level of supply. It can be argued that efficiency falls at higher intakes due to the ideal protein becoming progressively less ideal as more is supplied; or, in like vein, that at higher rates of protein retention, for some reason more protein is needed per unit of protein growth (or protein maintenance). Another plausible explanation is that at higher rates of protein retention, maintenance requirements increase due to some change in the characteristics of protein turnover, such as a reduction in the potential for amino acid recapture when the total volume of protein in the system becomes too great. A more comfortable and accommodating proposal is that the curvilinear response of Figure 4.6 is no more than a measure of the gradually increasing oversupply of protein above that which the animal can effectively use for protein retention. This last possibility probably does indeed account for the greater part of the fall-off in efficiency of use of ideal protein as intake increases (see also Fig. 4.5), but it begs the question as to what a pig's potential rate for protein growth might be and it certainly fails to be a totally adequate explanation for observed phenomena. If, for example, the requirement for maintenance is added to the picture (Fig. 4.6(b)), then the relationship between total supply and total usage still fails to reach 100 per cent efficiency of use (Fig. 4.6(c)) of absorbed ideal protein at all levels of intake apart from the very low. If it is to be assumed that at medium levels of intake protein is not supplied in excess, then it appears that there is some level of unavoidable mechanical loss, resulting in efficiencies of use of ideal protein not of 1.00 but of between 0.80 and 0.90 pertaining over the normal working range.

Figure 4.6 *Relationship found in practice between ideal protein supply and protein utilization in growing pigs: (a) protein retention alone; (b) together with maintenance requirement; (c) 100% efficiency, showing a 1:1 relationship between supply and use of ideal protein.*

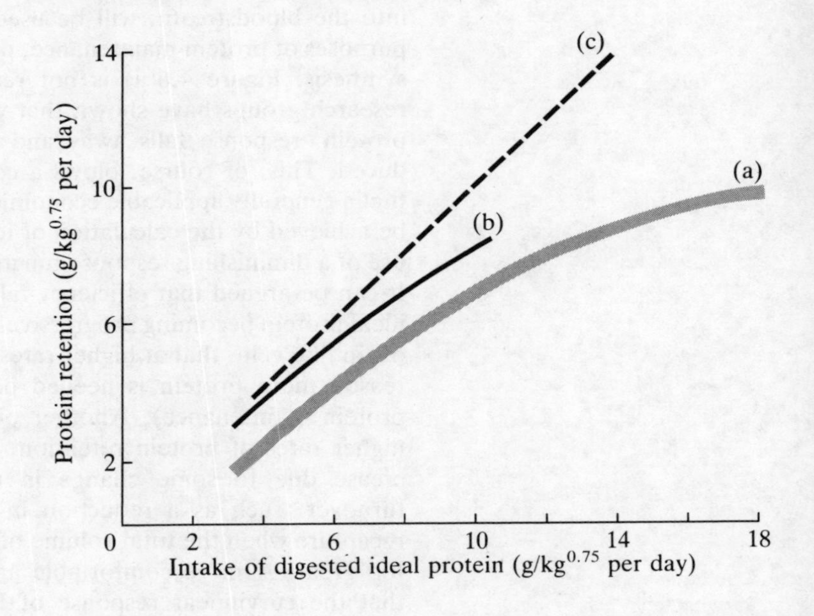

Fate of utilizable protein

Given that the diet has supplied *n* units of utilizable protein (absorbed protein of biological value 1; 'ideal protein'), the first call is for maintenance. In the growing animal, this is usually less than one-third of the total requirement. As maintenance protein simply replaces the losses due to the inefficiencies of protein recapture during protein turnover, then this maintenance allowance ends up deaminated in the urine.

Protein deposition in growing tissues or in milk has the next call on utilizable protein (Fig. 4.5). Where protein supply is less than the requirements of potential protein deposition and where there is enough energy available, then protein intake will limit protein growth and spare energy will go into fat production. Clearly, this is a grossly inefficient situation and normally protein would be supplied up to the requirement of the maximum rate of protein deposition.

Where protein supply is limiting protein deposition, all the ideal protein will be used up. The efficiency of use of ideal protein will be high and diminished only by 'mechanical' inefficiency and the needs of maintenance. The proportion of digested diet protein retained in the body will approach the determined biological value of the diet protein (the difference being maintenance (Fig. 4.5)). However, when protein is supplied in excess of that for which there is energy to match, or in excess of that needed to maximize potential protein deposition in growing tissues or in milk, then utilizable protein will overflow the system and be diverted to excretion in the urine. The *effective* biological value is in this way reduced and *utilizable* protein joins the *unutilizable* and is indistinguishable from it (Fig. 4.5). Distinction must be made therefore between the biological value of the diet protein and the biological value achieved by the pig (the latter being dependent upon the pig's ability and the energy supply, as well as the amino acids in the diet; the former only being dependent upon the amino acids in the diet).

In summary, the utilizable protein value of a mixed diet may be determined as the product of the CP content, the effective digestibility of the CP, the biological value of the protein and the extent, if any, of the inefficiency of use of ideal protein (mechanical inefficiency, often taken as 0.85 in the absence of more definitive information). Digestibility can be determined *in vivo*, or assumed to be about 0.75 if no extrinsic or intrinsic factors interfere. Biological value is best derived from chemical analysis for amino acids and then comparison with 'ideal protein'.

Next steps in feed evaluation

Equations for estimation of DE and the concept of ideal protein represent considerable steps forward in the effective evaluation of feed ingredients and mixed diets for pigs. Within the bounds of acceptable error, nutritive values may be derived and pig responses to foodstuff mixtures predicted. Perhaps then it is appropriate that the attentions of nutritionists should turn away from the conventional chemical axes of energy, protein, minerals and vitamins, and move next towards the more holistic and organic concepts, such as feed intake stimulation and nutritional welfare.

Limitations to feed intake curtail the efficiency of pig production over much of the life cycle; enhancement of feed intake would allow a quantum leap forward. The causal forces of variation in feed intake have hardly begun to be understood, but will certainly relate to aspects of animal behaviour as much as to the chemistry of food and the biochemistry of the pig's body. Without doubt, the quantitative definition of a feed ingredient will have to go beyond a statement of fibre, fat and amino acids, such as have been dealt with so far; it may be more important in relation to feed intake to know of the crop species and storage characteristics than chemical composition. Given the overridingly crucial influence of feed intake upon nutrient supply, it is a truism to point out that in many circumstances it is more important that a diet is readily consumed than that its chemical composition is especially superior.

Nutritional welfare is a complex concept and relates to the growing suspicion that pigs may be simultaneously adequately supplied with nutrients but poorly fed. The beneficial effects, for example, of fibre upon gut health and animal welfare are now becoming well appreciated. It is likely that many other aspects of the components of a diet will be found to be important elements in the process of alimentation. Can it be acceptable or nutritionally optimum for a pregnant sow to consume her total daily feed supply in only 10 minutes of her 24-hour day and to have filled no more than one-third of her stomach capacity in the process? In the natural state, animals have to work for their food and this most normal of all pursuits is still to be found to a greater or lesser extent in other domesticated livestock species. But for the pig, feeding has been reduced to a particularly bland and uninteresting activity, requiring only minimal mental and physical effort. Is it possible that production efficiency may be placed at a disadvantage by the methods currently employed to supply food to the animals? There is more to feeding pigs than providing for their nutritive requirement.

Chapter 5 Tactics and strategies for the nutrition of breeding sows

Introduction

The value of the reproducing sow is a function of her food costs to first conception, her pigs per year, her food consumption during breeding life and her worth when culled. Appallingly little is known of the nutritionally motivated causal forces of reproduction in sows. Experimentally, it is even difficult to accumulate adequate evidence as to the optimum weight and/or condition (fatness) of a sow for reproductive effectiveness. Science seems ignorant of the nutritional factors contributing to reproductive success, and therefore seeks blueprint dogmas for sow feeding and management. There is little understanding as to why any particular tenet might be effective and there is little confidence within the scientific community that different recommendations might not be found to be better, cheaper, easier or whatever. There is much fundamental nutritional research yet to be done with sows; in particular, to study the influence of fatness and fatty tissue change upon reproductive performance.

A new dimension to the sow feeding problem has followed from the advent of the hybrid gilt, and the greatly improved growth rates and leanness of finished pigs destined for slaughter. Much of the classical sow nutrition experimentation of Elsley and his colleagues used gilts with P2 backfat depths of 25–30 mm at the time of mating. The position now is that gilts are reaching mating weight much younger and carrying about half that amount of fat on their backs. This problem may not best be tackled after mating, but there has been little, if any, effective research and development work to see if the rearing of females destined for the breeding herd in a different manner to those destined for slaughter might not be a worth-while proposition.

Optimization of sow feeding strategies requires the resolution of the following, yet unresolved, issues:

- Which, if any, are the short-term influences of nutrition upon reproduction; and are these independent of long-term nutrition?
- What level of fatness is preferred by sows for successful conception, gestation and lactation?
- Given that fatty tissue catabolism is a normal part of lactation, then what are the optimum rates of lipid gains and losses through the reproductive cycle?

Progress over the last 25 years towards understanding the fundamentals of sow feeding has been erratic; and the nature of the scientific endeavour not so much inspired as dogged.

In the 1950s sows were fed to look fat; after all, the expectation for pigs was to be fat. Pig genotypes were considerably fatter at that time (30 mm of P2 fat at 100 kg would be normal: twice the late 1980s expectation), but breeding sows may also have been over-fat because their diet did not contain enough protein of adequate amino acid quality. Reproductive performance – as far as it could be judged from the records of the time – was not especially good at about 15 piglets per sow per year, even after taking account of weaning being at about 8 weeks post-partum.

Then, in the 1960s, flying in the face of accepted practice, the remarkable proposition was made by Lucas and Lodge from the Rowett Research Institute, and subsequently extended by Lodge at the University of Nottingham, that fatness may not be a positive attribute to reproductive performance and, in addition, unnecessarily costly in terms of food usage. A series of experiments seemed to show that no more than 2 kg/day of a conventional, cereal-based, diet was required by sows in pregnancy, this being not much more than about half the amount traditionally given. Potential savings of substantial quantities of sow food were evident and pig keepers immediately took to the new proposals. Where properly done, feeding regimes based on 2 kg per sow per day in pregnancy achieved a high degree of success – as they still do today. Unfortunately, in many circumstances the scheme failed, with serious consequences for the condition of the sows and for reproductive performance. Thin sow syndrome became commonplace.

Amongst the causes of thin sow syndrome were:

- Failure to increase protein, vitamin and mineral density in sow diets that were to be given in reduced amounts.
- Failure to eradicate intestinal worms from sows, the adverse effects of the worm burden being more serious in sows fed at low level.

- Failure to give sufficiently high levels of feed over the long and arduous 8-week lactation.
- Failure to account for farm-to-farm variation, many traditional units being entirely unsuitable for a feeding regime which 'sailed close to the wind'. Frequently, pigs were kept under much more adverse and rigorous circumstances than was the case in the research units upon which the '2 kg/day in pregnancy' regime was developed.
- Failure to account for variation amongst individual sows, on account of the erroneous assumption that all sows would react alike to a given feeding regime.
- Failure to provide each and every sow with its full 2 kg allowance.

Subsequently, a multiplicity of factors helped to ameliorate the symptoms of thin sow syndrome in pig herds. The industry lost a large number of small pig producers, who kept pigs only as a side-line. Weaning age came down from 8 to 6 weeks and then to 4 weeks to ease the lactation load. Parasite control became a part of normal pig-keeping routine. Diets were compounded to a higher density of protein and micronutrients, and designed on the assumption of a 2 kg pregnancy allowance. Sows came in from the cold to be given pregnant sow quarters, which reduced environmental stress to a minimum. Individual feeders, and then individual sow stalls, became common, thus ensuring each sow her fair share of feed. The 2 kg rule still remains the basis of pregnancy sow feeding throughout most parts of the world where pigs are kept intensively and given cereal/soya diets.

The 1960s and early 1970s saw a spate of opportunistic feeding schemes based upon the contention that sows could be given relatively little for most of the time, providing that they were fed relatively generously for short periods at crucial times. For example:

(a) Young gilts and weaned sows were to be flushed – often interpreted as given double – just before mating; this would increase the number of ova shed at oestrus and ensure a large litter. (This practice is still extant, not as a short-term 1-day wonder, but rather as a general need to feed all gilts generously, and to encourage high feed intakes between weaning and conception.)

(b) Generous feeding in early pregnancy would ensure a high rate of embryo implantation and survival (now discredited and the converse recommendation holding).

(c) A doubling of the feed allowance over the last 2–3 weeks of pregnancy would enhance foetal nutrition and ensure heavy piglets at birth (a practice that remains a matter for debate).

The most important development, however, was the evolution of the principle that sows were individuals characterized by variability of response to feeding regime. Because of variable response, generalized and dogmatic feeding rules are unlikely to result in optimum feeding practices. The late Professor Frank Elsley, after many years of painstaking work at the Rowett Research Institute, and through the uniquely innovative medium of internationally coordinated sow trials, derived a means of providing a plan for pregnancy feeding that would allow for between-sow variation. Frank Elsley's proposal was that sows should be given whatever it takes to gain a predetermined amount of live weight between one conception and the next (subsequently elucidated to be about 15 kg in each of the first 4 parities). At a stroke, this proposal provided pig-keepers with a general rule that would account for individuality of sow response. The principle is undoubtedly correct; although the preferred absolute rates of weight gain from parity to parity could be argued on account of the 15-kg standard implicitly assuming both: (a) that sows start their breeding life fat and should exploit these fat stores as they progress from one parity to the next, and (b) that the price of thin cull sows is approximately equal to the cost of young replacement gilts (thus rendering longevity an issue of no importance).

Elsley's proposals contained two flaws:

- Sow weighing – an essential prerequisite – has never been taken up by pig producers as being a practicable part of normal day-to-day pig-keeping routine.
- It is possible that sow weight gain may not be as important to the reproducing sow as the state of her body lipid reserves. Certainly, weight is an inadequate description of breeding potential, and itself fails to allow for between-sow differences in bone structure and mature size. (This is less of an oversight than may at first appear because, although total live weight is independent of total fatness, there is a much better relationship between live-weight gains and fat gains. That is, while light sows may be fat and heavy sows thin, a sow at any given weight and fatness subsequently gaining weight tends also to gain fat.)

The research target for the science of breeding sow nutrition should be the ability to predict pig responses to a change in regime. In order satisfactorily to simulate pig response, a great deal more information is yet required, regardless of whether the approach to system analysis be mechanistic (deductive) or empirical. The nub of the problem is that whilst nutrient responses may be observed and noted, the overriding influence of environmental variation upon the reproductive traits means that most experiments have been of inadequate scale to quantify reproductive response to nutritional manipulation.

To first conception

Young females destined for the breeding herd are invariably reared identically to, or even together with, commercial pigs grown for meat. This used to fit nicely with puberty at 160+ days and 90 kg, together with service at the third oestrus about 42 days later and 20 kg heavier, when the young gilts would contain about 30 per cent of body lipid (equivalent at this weight to about 30 mm or more of backfat at P2). After being taken out of the finishing pens at about 90 kg, but before mating, gilts used often to be 'hardened off' on a relatively restricted feeding regime. It may perhaps have been to counteract the adverse consequences of this parsimonious feeding regime being used during the run-up to first conception that the idea of 'flushing' arose, receiving special attention from workers at the University of Nottingham. After a series of experiments proving, and then disproving, that high levels of feeding for short periods prior to ovulation could increase the number of piglets born in the litter, current understanding seems to be that benefit can be gained in terms of numbers born from high levels of feeding in the 3 weeks prior to mating; that is to say, between the first and the second or the second and third post-pubertal oestrus. However, it remains unclear whether this is independent of, or merely part of, the creation of a satisfactory general body condition status. In retrospect, it has not been easy to find scientific evidence to follow up the original work of the 1960s from the USA and to show that the ovulation increase with each oestrus following puberty has any firm relationship with numbers born at the end of the pregnancy.

Since the pioneering work of the 1960s and early 1970s, the type of animals coming forward into breeding herds has dramatically changed. Maiden gilts, far from being plump and fat, tend to have an appearance no different from a modern top-grade bacon pig and carry very low levels of fat in consequence (Fig. 5.1). Modern improved strains of pigs can reach 90 kg at 140 days or less with little difficulty, and it is likely that this trend of increasing growth rates, decreasing age and decreasing fatness will – if anything – accelerate. The animal that is now being taken out of the finishing

Figure 5.1 ***Modern hybrid strains of pigs (b) are much less fat than their predecessors (a).***

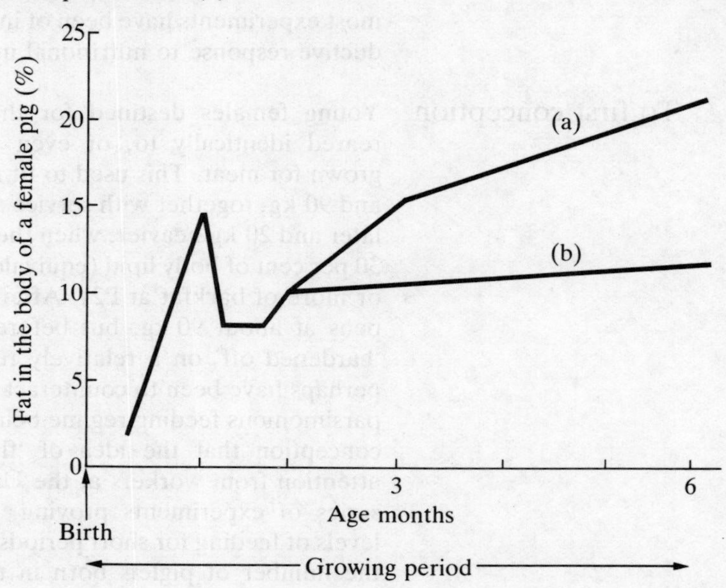

pens to be used as a replacement female for the breeding herd is therefore very different from what was the case only 15 years ago.

Present diagnosis is that improved strains of fast-growing lean pigs come up to what would usually be considered as an appropriate mating weight (105 kg) too young and too thin. This problem can never be resolved in conventional terms because merely letting time go by before first mating carries with it the danger of the pigs being grown slowly, and therefore becoming even thinner and quite unacceptably lean at the beginning of their breeding life. The solution seems to be to feed high levels, in order to encourage fatness. This, however, now incurs the considerable expense of additional feed, as well as of additional time (Fig. 5.2). Besides, as fast lean growth is associated with a greater mature size, it should not be assumed that the extra growth achieved will necessarily be of fatty tissues.

In the short-term, generous feeding to allow at least 750 g daily live-weight gain from 90 kg up to mating appears to be the appropriate tactic. Such gains are likely to follow from a feed

Figure 5.2 *If a target P2 backfat depth of, say, 25 mm is set, then maiden gilts with 16 mm of fat at 90 kg will require to be given high levels of feed, and to gain appreciable live weight, before the target is met. It is evident that, even with generous feeding, a realistic expectation of fatness at 115 kg live weight is no more than 20 mm P2.*

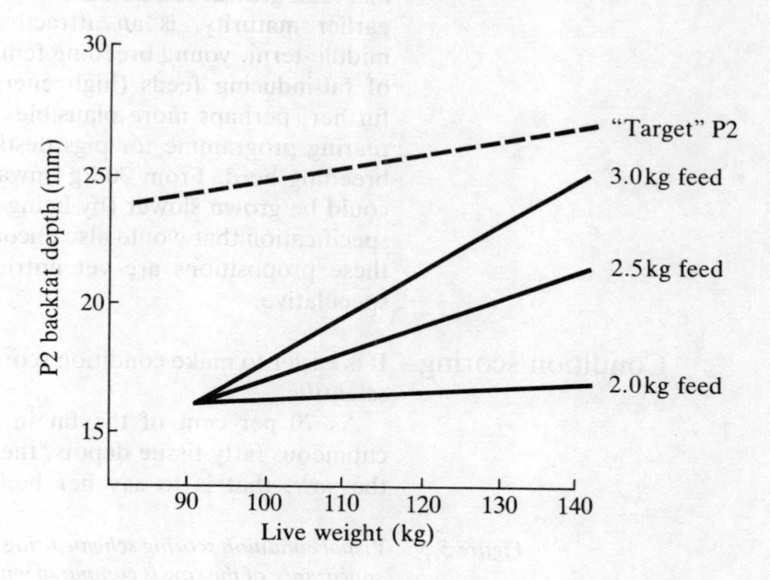

allowance of about 2.5 kg of a high-quality cereal/soya-based diet with 165 g crude protein (CP) per kilogram. In hard environmental conditions, a more generous feeding level may be needed. Field experience has led to the observation that benefit can accrue from leaving young potential breeding females until they are 115–120 kg before putting them with the boar for first mating. In the longer term, however, new strategies, as yet unresearched, will be required.

The targets for these new strategies must be:

- A small breeding female, which will produce annually at least 28 large, fast-growing offspring.
- A female achieving puberty at an early age and coincidentally with a reasonable level of fatness.

The small size will ensure minimum feed usage in the breeding population, while a certain level of fatness provides some nutritional insurance against unforeseen variations in feed supply or in health. The sooner puberty is reached, the lower will be the cost of production of replacement females.

Some of these ends could be achieved by nutritional or genetic means. Breeding specialized dam lines, particularly if prolificacy and lean growth can be maintained in the face of smaller size and earlier maturity, is an attractive long-term prospect. For the middle-term, young breeding females could be offered high levels of fat-inducing feeds (high energy/low protein) from 90 kg. A further, perhaps more plausible, possibility would be a selective rearing programme for pigs destined to be replacements for the breeding herd. From 30 kg onwards, potential breeding females could be grown slower (by being fed less) and given a nutritional specification that would also encourage them to become fatter. As these propositions are yet untried, their outcome must remain speculative.

Condition scoring

It is easier to make condition scoring objective than it is to make it scientific.

As 70 per cent of the fat in pigs may be found in the subcutaneous fatty tissue depots, then it follows that the fat status of the sow, that is to say her body condition, can be judged by

Figure 5.3 *Visual condition scoring scheme using a diagrammatic standard. The appearance of the sow is compared with the picture and the animal given a score appropriately.*

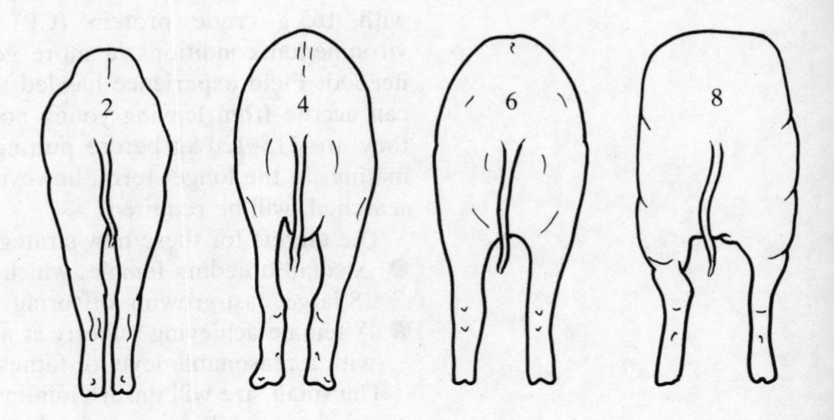

appraisal of her external appearance. A visual assessment may be compared to a photographic or diagrammatic standard (Fig. 5.3), or a physical assessment may be made by manual palpation of those of her body parts that are especially sensitive to fat changes, such as the areas over the spinal processes.

Visual condition score (CS; using a 10-point scale) is well related to P2 backfat depth as measured with an ultrasonic probe (USP2):

$$USP2 = 2.9\,CS - 0.7$$

The use of sow fatness (through the medium of a condition scoring technique) as a monitor of nutritional adequacy for breeding proficiency has the great advantage of being independent of feed level, health, environment and management; condition score being the *outcome* of the interactions of all these various forces impinging upon the sow.

It may be assumed that the sow is likely to perceive food, at least in part, as a means to a reproductive end. But the science of reproductive physiology is not adequate for a quantitative understanding of the relationships between nutrient supply and breeding effectiveness. In these uncertain circumstances it is perhaps justified to take the view that fatness may be a reasonable indicator of propensity to reproduce.

Target proposals for feeding strategies made in the form of condition scores surmount the problem of sows in different environments reacting in a variable way to feed level. The condition score target demands that the animal be fed what it takes – not fed a certain amount. In consequence, a sow judged too thin but already on a high feed level will require to be given more, whilst one too fat already on a low feed level will require to be given even less. The problem of the 2 kg/day in pregnancy rule being reasonable for a sow in a warm environment but unreasonable in a cold one, is removed by shifting the rule on from the level of cause up to the higher level of effect.

Usually, a condition score of above average (6/10) is targeted for the end of pregnancy, and one below average (4/10) accepted at the end of lactation. Under normal circumstances these targets can be achieved by relatively minor adjustments to feed levels. Sows thinner than 4 or fatter than 6 merit special attention.

An example of an objectivized feeding procedure for use on breeding units employing condition scoring techniques is given in Table 5.1. This procedure has not been taken up with enthusiasm by

Table 5.1 Feeding sows according to body condition

1. Using all the sows in the breeding herd that are pregnant or empty (that is, all sows not lactating), determine the average daily feed allowance per sow. Record this number.

2. Determine the condition score of all the sows in the herd that are pregnant or empty. Score according to the photographs and do not score using less than whole numbers. The target average condition score for the whole herd should be about 5. Usually, young sows should have a condition score of rather more than 5, while old sows should usually be rather less than 5. Newly weaned sows should score about 4, while sows due to farrow should score around 6.

3. If the average condition score for all pregnant and empty sows in the herd is above 5, then the average daily feed allowance (as per paragraph 1) should be reduced. If the average condition score is less than 5, then the average daily feed allowance for the sows should be increased. In the first event, increases and decreases in the average daily feed allowance should not be greater than plus or minus 0.2 kg per sow per day.

4. Although the average condition might be satisfactory, this average can be made up of sows that are thin and sows that are fat. To deal with this, individual animals must be fed according to their individual condition scores.

5. For each individual sow, determine the condition score immediately after weaning the piglets. Feed according to the Table below until a condition score of 6 is reached for the particular individual in question. Upon reaching condition score 6, the sow should be returned to the average daily feeding level (as per paragraphs 1 and 3).

6. At weaning the sow should be changed immediately from her lactation feed allowance to a suitable regime to prepare her for mating. Feed allowances between weaning and conception are likely to involve feeding at least 2.5 kg of lactating sow diet and in many instances feeding to appetite could be considered. (Sows will usually eat about 3.5 kg or so and it is unnecessary to present more than 3.5 kg of food daily.) Sows failing to be mated within 10 days of weaning should be fed until they are pregnant according to the Table below.

7. In lactation, sows may be fed to appetite; according to the size of the litter; or to a high fixed level. In general, high-level feeding is justified in lactation, in order to avoid an excessive rate of fat loss and thinness at weaning.

Increases or decreases to the average daily feed allowance for pregnant sows consequent upon their individual body condition score. These feed allowances relate only to the pregnancy period after mating.

Sow body condition score	Likely feed requirement in addition to the average daily feed allowance for sows in pregnancy (kg/day)
2	+ 0.6
3	+ 0.4
4	+ 0.3
5	+ 0.2
6	Average daily feed allowance
7	− 0.2
8	− 0.3

practical pig producers and is often considered to be an unnecessary embroidery. Many professional unit managers now work on the basis that, if the sow is thin, she should be fed more; if very thin, much more. The definition of 'more' is unit-dependent and learned by experience. The most important criterion is the target condition score chosen as the unit standard but, unfortunately, score 6 can be fatter on some units (looking more like 7) than on others (where they can look more like 5).

The problems with condition scoring systems are:

1. Condition score measures fatness only indirectly, and can be prone to error consequent upon variation in sow size and shape.
2. Condition score is dependent upon the subjective opinion of the scorer and standards can drift.
3. Target condition scores for various points in the reproductive cycle lack scientific validation. While individuals may have views about adequate condition states through the reproductive cycle, the relationships between condition score and reproductive performance have yet to be quantified in science.
4. The response of condition score to change in feed intake is not well documented and variable.

Given that it is unlikely that commercial managers will wish regularly to weigh their breeding sows, it is reasonable to assume that condition scoring will become the way in which practical pig producers monitor the success of their feeding strategies and decide upon appropriate adjustments to their feeding tactics. Solution to the four problems listed above would seem worthy of the considerable scientific attention they are now receiving.

Most research programmes have returned to the basic assumption behind condition scoring philosophy: that level of fatness controls reproductive proficiency. Subcutaneous fatness may be conveniently measured objectively by determining with an ultrasonic probe the depth of fat at the P2 site. One mm change in P2 is about equivalent to a 1 percentage unit change in the fat composition of the live weight:

$$\text{Body fat (\%)} \simeq 1.0 \times \text{P2 (mm)}$$

Live-weight and fatness changes

Sows gain body weight in pregnancy and lose weight during lactation (Fig. 5.4). In addition, at parturition, they drop the combined weight of the foetuses at term plus about another 50 per cent of placenta and fluids.

Figure 5.4 *Live-weight changes in breeding sows.*

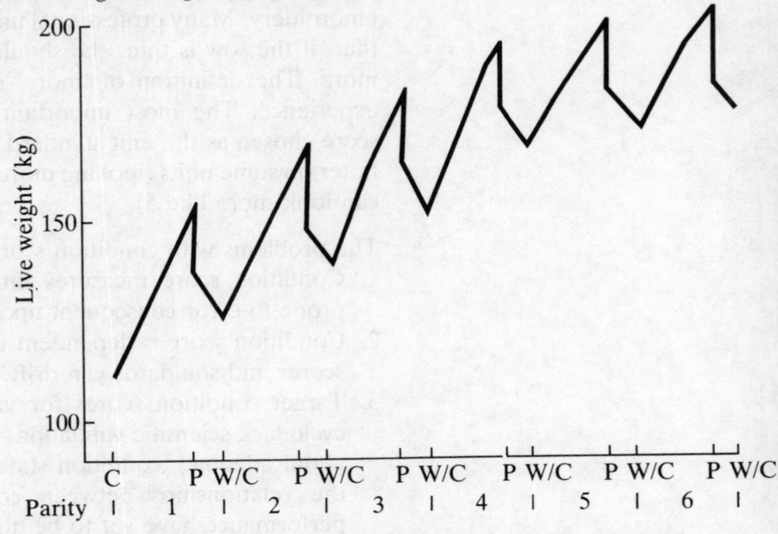

C = conception; P = parturition; W/C = weaning/conception.

The greater the feed allowance in pregnancy, the greater the pregnancy gains. The less the feed consumed in lactation, the greater the lactation losses. Fig. 5.4 shows the original Elsley proposals of 12–15 kg of gain, parity upon parity, up to the fourth pregnancy. During the early parities the sow will be aiming to achieve her mature lean mass. It may thus be assumed that pregnancy gains are of both lean and fat. Lactation losses are most likely to be fat alone, as this is the prime substrate needed for milk production. *In extremis*, there may be some lactation losses of protein, but this would only occur if the diet was protein deficient or the sow had expended all her fat reserves. (A phenomenon always noted on nitrogen-balance trials, and of at least passing academic interest, is the loss of urinary nitrogen in the first 2 days after parturition. The reason/cause of this is not clear, but it may be due to some labile protein being associated with the most labile fat stores that are kicked into activity by the hormonal events attending parturition.)

For many years there has been a discrepancy concerning the assumed contribution of live-weight loss to the energy economy of lactating pigs and dairy cows. On the basis of losses being fat alone, a value of around 46 MJ dietary metabolizable energy

(ME) equivalent can be ascribed to 1 kg of weight loss from the back of a lactating pig; while on the basis of body-weight losses being of similar composition to body-weight gains, a value of around 28 MJ dietary ME equivalent has been ascribed to 1 kg of weight loss from the back of a dairy cow. The calculation may be made thus:

The energy in 1 kg of body fatty tissue is about 38 MJ.
The efficiency of conversion to milk is about 0.85, which gives 32 MJ milk energy.
32 MJ of milk energy requires 46 MJ diet ME if the efficiency of diet ME to milk energy is 0.7.

Unfortunately, all such calculations are highly tentative because it should neither be assumed (a) that weight losses are only of fatty tissues, nor (b) that 1 kg of fatty tissue loss is necessarily reflected by 1 kg of weight loss. It is now evident that in all mammals the catabolism of lipid from tissues can take place simultaneously with replacement into those tissues of water. The water:protein ratio in the body can change from 3.5 to 4.0 as lactation progresses. Because of this compensatory effect, measured weight losses fail to reflect lipid use and the energy yield per kilogram of live-weight loss can be considerably in excess of 50 MJ of dietary ME equivalent.

Work with improved strains of pigs has shown unequivocally how weight gains in the first two parities take place commensurately with significant fat losses (Fig. 5.5).

The clear demonstration that breeding sows can gain weight and lose fat simultaneously places a new dimension on the biology of nutrition and reproduction. First live weight and live-weight change cannot be used as a general indicator of body condition nor of ability to breed. Secondly, it seems as if pigs can simultaneously increase lean tissue in the course of satisfying their goal of mature lean mass, while also decreasing fatty tissue as a means of supplying nutrient substrates for milk production. Figure 5.5 shows how in pregnancy both weight and fat are gained, but more weight than fat; whilst in lactation both weight and fat are lost, but more fat than weight. Gains of 11 kg live weight in the course of a parity appear to have taken place despite absolute losses of about 4 kg of fat.

The responses shown in Fig. 5.5 lead to the conclusion that sow weight gains and fat losses are natural expectations as breeding life

Figure 5.5 *Changes in fat depths and live weights of sows over two parities. (From Whittemore, C. T., Franklin, M. F. and Pearce, B. S. (1980.)* Animal Production **31***: 183)*

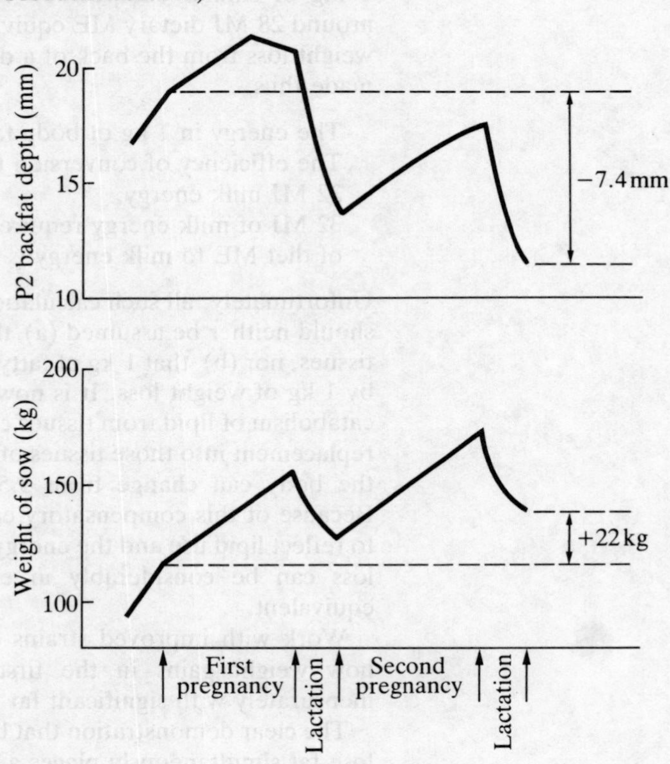

progresses. But it is *not* to be assumed that these gains and losses are desired or that they should be willingly accepted at the levels shown. The pigs concerned were modern hybrids, but fed at the rate of 1.8 kg/day in the first pregnancy and 2.3 kg/day in the second, and the average lactation feed allowance was 4.8 kg/day. It is evident that, by gaining mostly lean in pregnancy and losing mostly fat in lactation, these animals are running out of fat at a dramatic rate, which evidently cannot long be sustained.

Figure 5.6 shows how the distributions of P2 backfat depth throughout the population of pigs on the same trial altered from entry at about 90 kg through to weaning at parity 2. If it is

Figure 5.6 *Distribution of fatness amongst breeding pigs. Nine populations of pigs, located at one centre, are represented. All pigs were treated alike. A particularly lean population was designated C and a particularly fat population, H. Mean fatnesses for these two extremes are shown in the diagram.*

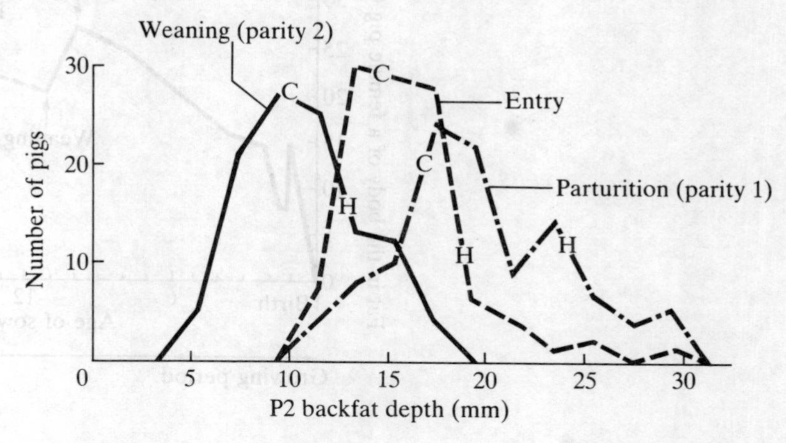

accepted that sows are likely to show re-breeding difficulties at P2 fat depths of below 10 mm, then a large proportion of these sows were clearly at risk.

Given young breeding females with about 20 per cent of fat at mating and about 25 per cent fat at parturition, the conventional pattern of weight gains (Fig. 5.4) and fat losses (Fig. 5.7) may be pursued with confidence as a proper nutritive strategy for sow feeding. However, the newer learner hybrid strains have fat percentages nearer 12 per cent at mating and 15 per cent at parturition, so the pursuit of a fat-depleting regime will be likely to lead to disaster (Fig. 5.8).

If it is supposed for the time being, and possibly for the foreseeable future, that young breeding sows will come to first mating lean, then it follows that sow feeding strategies must maintain fatness throughout breeding life (Fig. 5.9). This means that lactation fat losses must be discouraged and minimized, while pregnancy fat gains must be encouraged within the bounds of the requirements of high embryo survival.

These realizations led to a series of experiments at Edinburgh, Nottingham, Shinfield, the Rowett Research Institute, Australia and elsewhere. The objective in each case was to try to understand the

Figure 5.7 *Conventional pattern of fat losses in sows throughout breeding life. Note the generous levels of fat present at the start of the breeding period.*

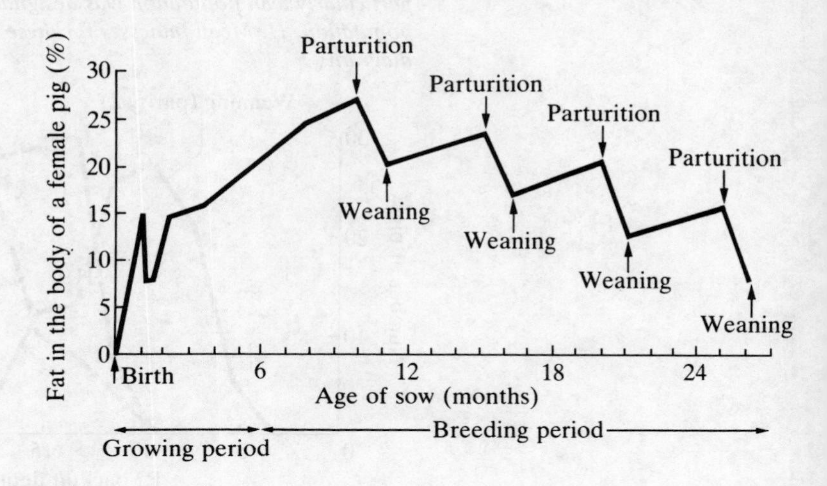

Figure 5.8 *The broken line (b) shows the likely consequences of imposing a conventional feeding regime designed for sows of type (a) upon modern improved hybrid strains, which start their breeding life much less fat. The dotted line (c) indicates minimum fatness for effective breeding.*

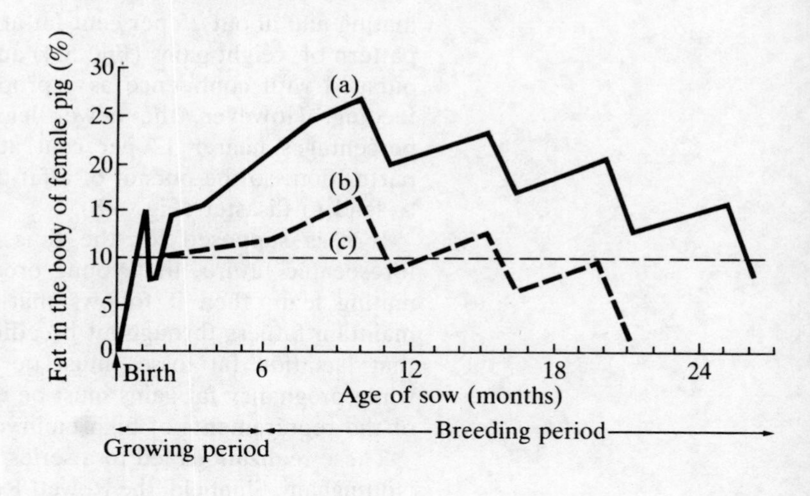

Figure 5.9 *The broken line (b) represents the likely preferred pattern of fat change in breeding sows starting their breeding life with limited fat stores. In comparison with the conventional picture (a), sows of type (b) must be discouraged from losing much fat in lactation. The dotted line (c) indicates minimum fatness for effective breeding.*

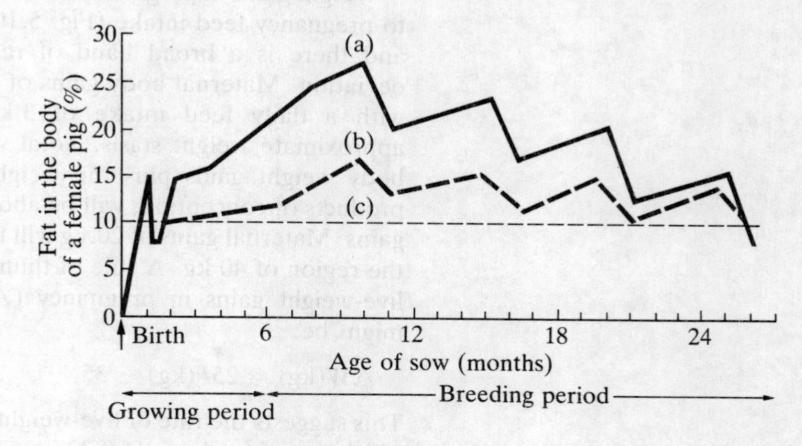

relationships between weight change, fat change and reproductive performance. The picture that begins to emerge is yet unclear; it may, however, serve to give an impression of some of the biological occurrences surrounding the interaction of nutrition and reproduction.

Feeding in pregnancy

Pregnant animals tend to eat more than when non-pregnant; but even on the same feed intake, pregnant animals have higher rates of nitrogen retention and live-weight gains. The efficiency of use of nitrogen for protein retention can rise from 0.45 to 0.60, with the consequence that the litter and other products of conception can be created for no extra dietary protein supply. These positive, anabolic, benefits to the maternal body are lost soon after parturition when there is a flush of nitrogen in the urine, the final balance showing pregnancy to have resulted in a small net loss of maternal body protein.

Mammary development occurs almost entirely during pregnancy. Fatty tissue deposition appears to be accelerated during the first three-quarters of pregnancy but in the last quarter it is often the case that some fat may be catabolized as the mother goes into negative energy balance. Whether this loss of fat, even during pregnancy, is an unavoidable biological phenomenon, or the result of underfeeding in late pregnancy, is not understood. The picture of fat loss from the 82nd day of pregnancy is particularly evident in young gilts (see Fig. 5.5).

Weight gains in pregnancy are, not unexpectedly, closely related to pregnancy feed intake (Fig. 5.10). Response is highly variable and there is a broad band of response with a large standard deviation. Maternal body gains of 40 kg or so may be associated with a daily feed intake of 3 kg, while 1.5 kg will achieve approximate weight stasis. Total weight gains, that is, maternal body weight gains plus the weight of the litter and the other products of conception, will be about 20 kg greater than maternal gains. Maternal gains of 20 kg will therefore relate to total gains in the region of 40 kg. A rule of thumb regression relating maternal live-weight gains in pregnancy ($\triangle W$) to daily feed intake (F) might be:

$$\triangle W(\text{kg}) \simeq 25F(\text{kg}) - 35$$

This suggests the rate of live-weight gain may be increased with an efficiency of feed use of 0.22 or a feed conversion rate of 4.5:1.

Figure 5.10 *Response of maternal live-weight gain to feed intake in pregnancy. The diet is assumed to contain about 13 MJ digestible energy (DE) per kilogram.*

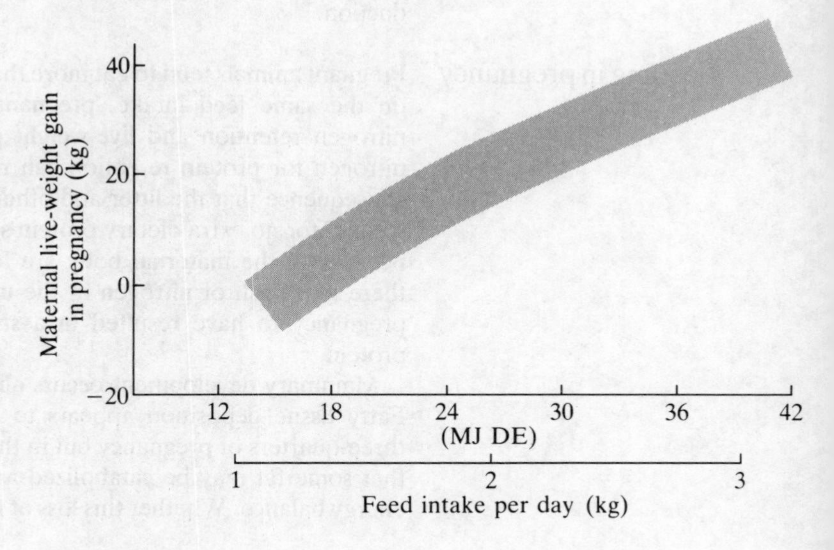

The relationship between pregnancy feed intake (F) and maternal fatty tissue gains (ΔT) is shown in Fig. 5.11. The corresponding rule of thumb regression is:

$$\Delta T(\text{kg}) \simeq 15F(\text{kg}) - 30$$

The broad approximations contained within these two regression coefficients imply that increasing weight gain in pregnant sows is usually achieved largely, but not entirely, by the deposition of fatty tissue and that weight gains can proceed at lower feed intakes than will support fatty tissue gains. A daily intake of 2 kg of an average diet (containing 13 MJ digestible energy (DE) per kilogram as given) seems to be associated with maternal weight gains of about 15 kg, total body weight gains of about 35 kg and little gain of fatty tissue.

In a recent Edinburgh experiment, a feed level of 2.3 kg* daily in pregnancy gave average pregnancy maternal body-weight gains of about 15 kg, while a feed level of 1.7 kg/day gave maternal gains of about 5 kg. These were associated with P2 fat changes in

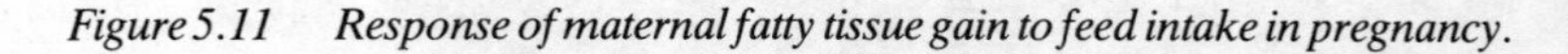

Figure 5.11 *Response of maternal fatty tissue gain to feed intake in pregnancy.*

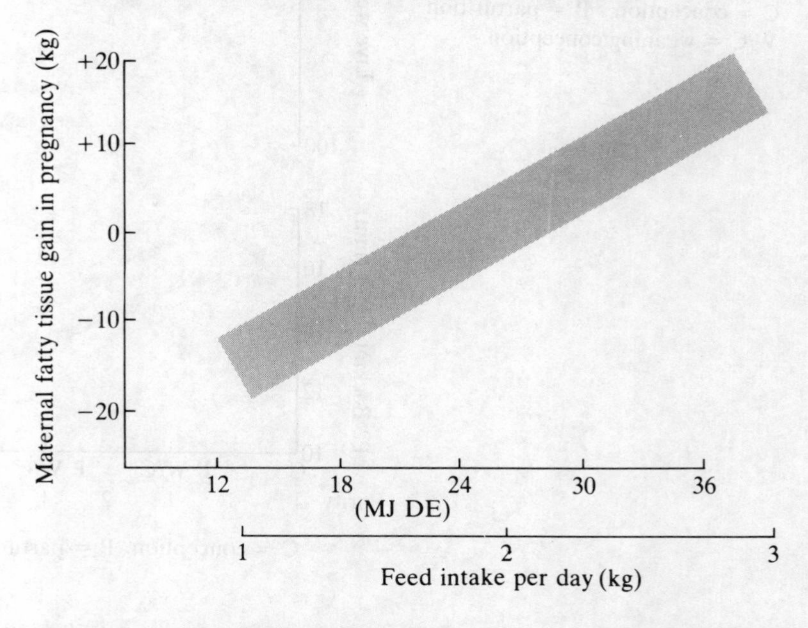

*In all these trials the feed used was a high-quality cereal/soya mix of 13 MJ DE and 165 g CP per kilogram/air-dry diet.

the region of +3 and −0.5 mm respectively. An additional 500 kg of food daily in pregnancy can be expected to give an additional 100 g of daily live-weight gain and 0.03 mm of daily P2 increase. Overall, weaning-to-weaning, responses to pregnancy feeding were less than this because sows gaining most in pregnancy invariably lost most in lactation (Fig. 5.12). This response was more marked with fat than with weight. Consequently, the two feeding levels in pregnancy produced greater differences in live weight than in fatness (Fig. 5.13). Indeed, both groups of pigs seemed to be targeting for similar fatness levels. This was achieved by the smaller, thinner sows adjusting their milk yields downwards and thus reducing the resultant rate of growth of the sucking piglets:

Figure 5.12 Sows gaining more live weight or more fat during pregnancy also tend to lose more during lactation. Following especially low weight gains in pregnancy, it is even possible for sows to gain weight during lactation. It would be most unusual for sows to gain fat in lactation.

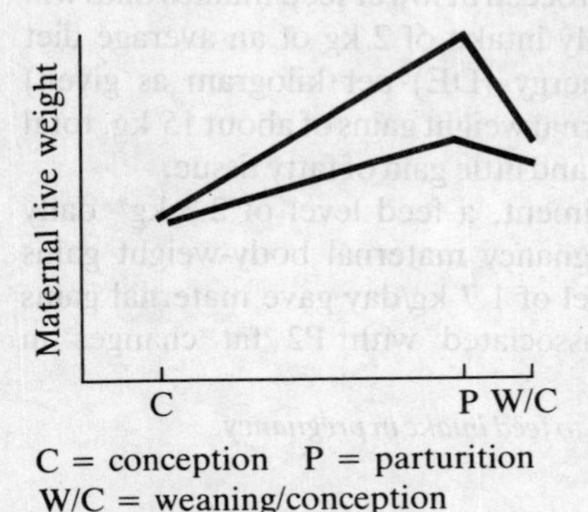

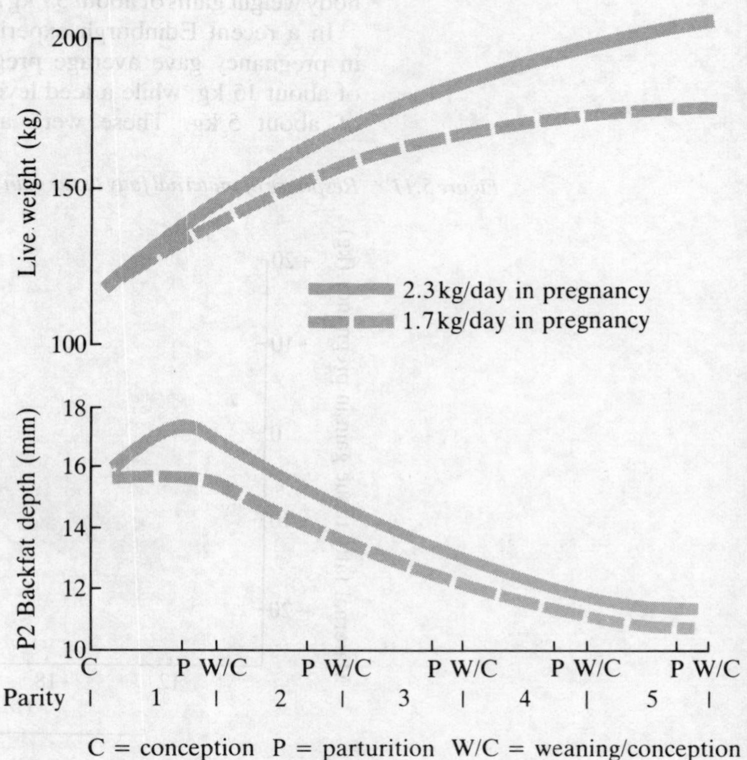

Figure 5.13 Influence of two feeding rates in pregnancy upon live weight and fatness of breeding sows over five parities.

Figure 5.14 *Influence of two feeding rates in pregnancy upon days empty and total litter weight. Sows given the higher rate weaned heavier piglets and reproduced more effectively.*

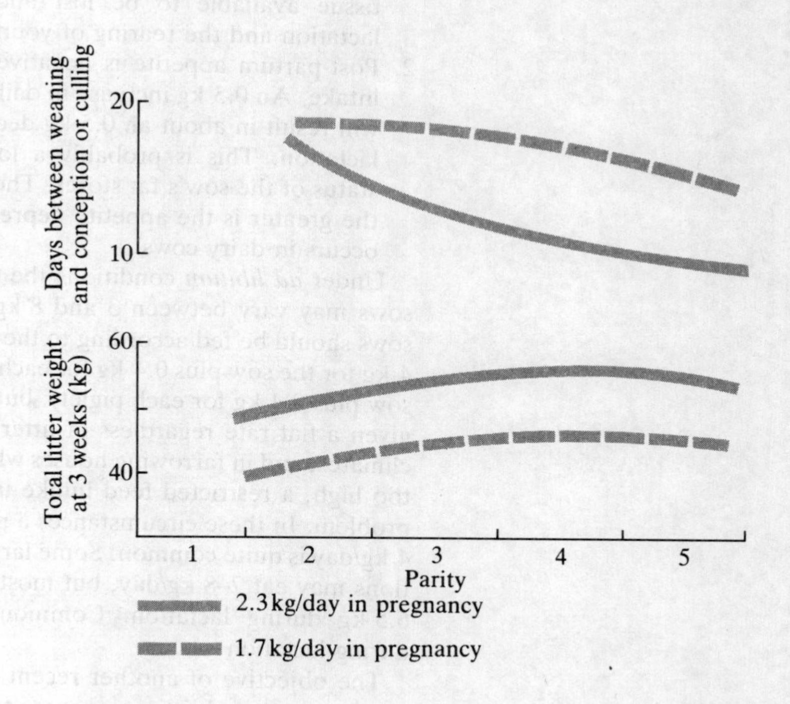

Weight of litter (kg) at 28-day weaning = 57 + 1.5 sow fat loss (mm P2) + 0.46 sow weight loss (kg).

The lighter, thinner sows also helped to adjust their state by delaying re-breeding, and increasing the days between weaning and next conception. Days between weaning and conception were negatively related to both sow fatness and sow live weight at weaning: the fatter, heavier sows re-bred sooner after weaning (Fig. 5.14). At the fifth parity, sows fed 2.3 kg in pregnancy were 20 kg heavier than those given 1.7 kg, but less than 1 mm fatter. More than 20 kg of extra litter weight was weaned at 28 days post- partum. Culling rates were higher for the low-fed group (60 *v* 35% by parity 5).

Feeding in lactation

Sow weight and fat losses in lactation are related to weight and fat gains in pregnancy. The causes are two-fold:

1. Sows of greater weight and fatness at parturition have more tissue available to be lost and to be volunteered toward lactation and the rearing of young.
2. Post-partum appetite is negatively related to pregnancy feed intake. An 0.5 kg increase in daily allowance during pregnancy will result in about an 0.5 kg decrease in daily appetite during lactation. This is probably a long-term effect linked to the status of the sow's fat stores. The fatter the sow at parturition, the greater is the appetite depression. A similar phenonemon occurs in dairy cows.

Under *ad libitum* conditions the daily feed intakes of lactating sows may vary between 3 and 8 kg daily. It is often stated that sows should be fed according to the size of the litter (for example, 4 kg for the sow plus 0.4 kg for each piglet over six; or 2 kg for the sow plus 0.4 kg for each piglet), but most often lactating sows are given a flat rate regardless of litter size. In many Mediterranean climates, and in farrowing houses whose ambient temperatures are too high, a restricted feed intake in lactation presents a genuine problem. In these circumstances a maximum feed consumption of 4 kg/day is quite common. Some large sows kept under cool conditions may eat 7–8 kg/day, but most sows' appetites rarely exceed 6.5 kg during lactation. Commonly, sows eat about 5 kg/day during lactation.

The objective of another recent Edinburgh experiment was to measure sow and litter responses to level of feeding in lactation, and assess the consequences in terms of weight and fatness changes. Four levels of feed† were offered during the 28-day lactation: 2.0, 3.5, 5.0 and 6.5 kg/day. These levels were chosen to be consistent with the range of feeding levels found in practice. Lactation weight losses were directly related to feed allowance (Fig. 5.15). At the highest feed intakes, maternal body-weight gains of more than 10 kg were achieved in the course of the 28-day lactation, but at the lowest intakes some 30 kg of sow maternal body weight was lost. Fat losses as measured by the change in ultrasonic P2 measurement show similiar trends: the less the feed, the greater the rate of fat loss (Fig. 5.16).

†A high-quality cereal/soya mix of 13 MJ DE and 165 g CP per kilogram/air-dry diet was employed.

Figure 5.15 *Influence of feed intake in lactation on weight losses in sows.*

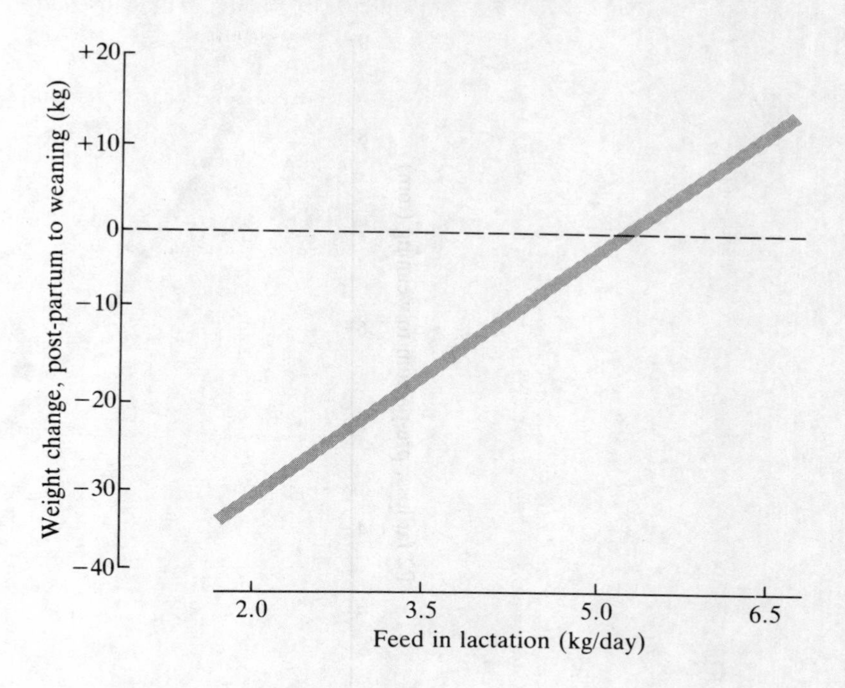

Linear regression analysis gave:

Weight loss during 28-day lactation (kg) = 50.0 − 9.8 lactation feed intake (kg/day)

Fat loss during 28-day lactation (mm P2) = 11.3 − 1.4 lactation feed intake (kg/day)

These equations suggest:

- Lactation weight loss may be largely prevented at feed intakes of above 5 kg/day, which represents an entirely feasible opportunity.
- Lactation fat loss may be largely prevented at feed intakes of above 8 kg/day, which represents an entirely unfeasible opportunity.
- An extra kilogram of food over a 28-day lactation will save about 10 kg of maternal body-weight loss and about 2 kg of maternal fatty tissue loss.

Figure 5.16 Influence of feed intake in lactation on fat losses in sows.

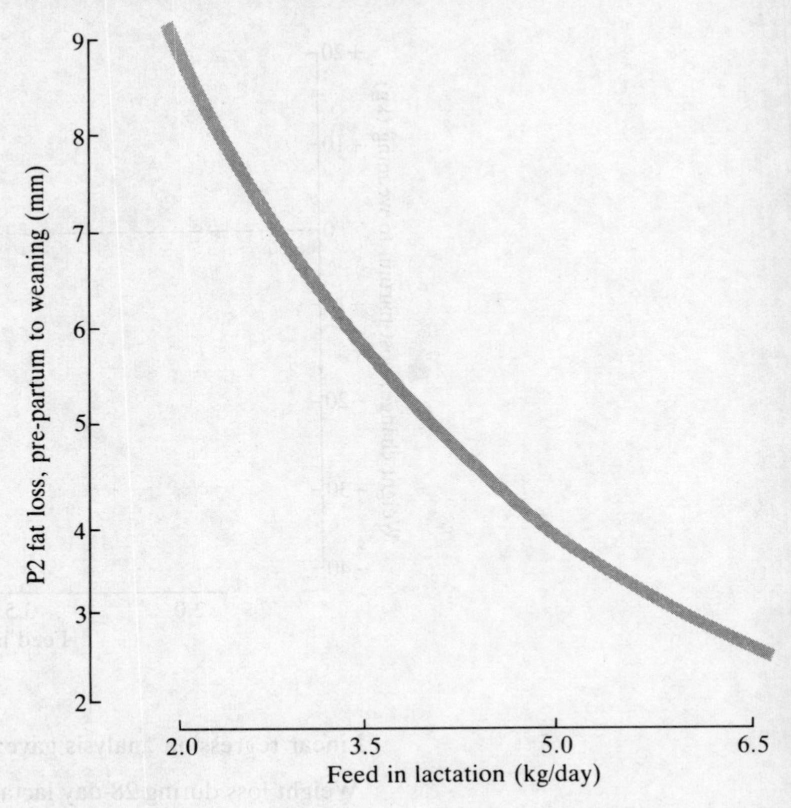

Feed intake in lactation had a positive effect upon milk yield and the weight of piglets at weaning (Fig. 5.17):

> Weight of piglet at 28-day weaning (kg) = 5.3 + 0.22 lactation feed intake (kg/day)

Most illuminating of all, however, was the finding that between 4 and 18 kg of lipid was lost by sows during the 28-day lactation, with the consequent relationship:

> Fat loss (kg) during 28-day lactation = 7.5 + 0.3 live-weight loss (g).

The equation shows how, even at weight stasis (zero weight loss), there was nevertheless some 7.5 kg of fat lost. This equation

Figure 5.17 Influence of feed intake in lactation on weight of weaned litter.

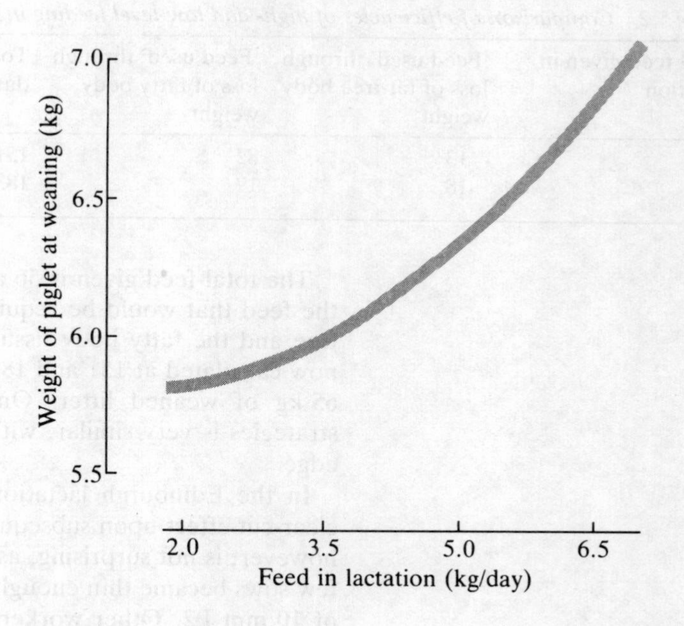

demonstrates yet again the independence of changes in weight from changes in fatness in lactating females and also the powerfulness of the drive that lactating mammals have to lose fatty tissue during lactation. It is unlikely that lactation feeding strategies can ever avoid the loss of fatty tissue. Mammalian body condition changes anticipate lactation fat loss by encouraging pregnancy fat gain. Seal pups may grow 1.5 kg daily on 50 per cent fat milk while the mother eats nothing.

As with the pregnancy feeding experiment, piglet weaning weight was also positively related to the rate of fat loss from sows in lactation.

It is germane to examine the comparative efficiencies of high- or low-level lactation feeding. In Table 5.2, the two extremes of 2 and 6.5 kg/day over the 28-day lactation are compared.

Table 5.2 Comparison of efficiencies of high- and low-level feeding in lactation

Total feed given in lactation	'Feed used' through loss of fat-free body weight	'Feed used' through loss of fatty body weight	Total 'feed used' during lactation	Total weight of litter produced
56	13	82	151	50
182	−18	19	183	65

The total feed given is 56 and 182 kg. To this should be added the feed that would be required subsequently to replace the fat-free and the fatty body tissue losses. Total feed actually used is now calculated at 151 and 183 kg, to produce respectively 50 and 65 kg of weaned litter. On this basis the efficiency of both strategies is very similar, with the high feed level having a slight edge.

In the Edinburgh lactation feeding experiment there was no clear-cut effect upon subsequent reproductive performance. This, however, is not surprising, as the treatments were short-term and few sows became thin enough to reach the purported danger level of 10 mm P2. Other workers have, however, proposed that for each 1 per cent of fat lost in lactation there will be 0.1 pigs less born in the next litter, to which could be added the view of many stockmen that, for thin sows, delaying conception by 21 days can give an extra pig per litter. And, Dr Ian Williams is unequivocal in his support of evidence from a variety of sources, mostly Australian, that the greater the weight and fatty tissue losses during lactation, then the longer will be the weaning to conception interval. Figure 5.18 represents these propositions. As body-weight losses in lactation increase, then so does the weaning to successful mating (conception) interval. Average intervals are usually about 10 to 12 days(‡), so re-breeding problems seem to be likely to become evident with maternal body-weight losses in lactation of greater than 12 kg for sows with little fat, or 30 kg for sows with ample fat. The implication

‡Sows normally show oestrus 3–5 days after weaning; the average results from about 20% of sows returning.

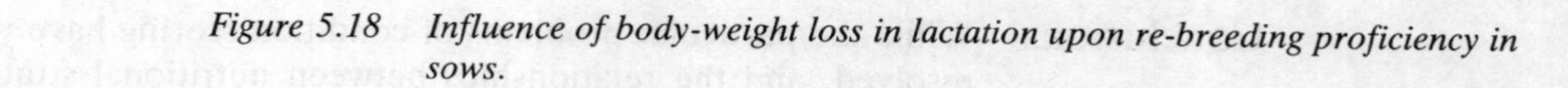

Figure 5.18 *Influence of body-weight loss in lactation upon re-breeding proficiency in sows.*

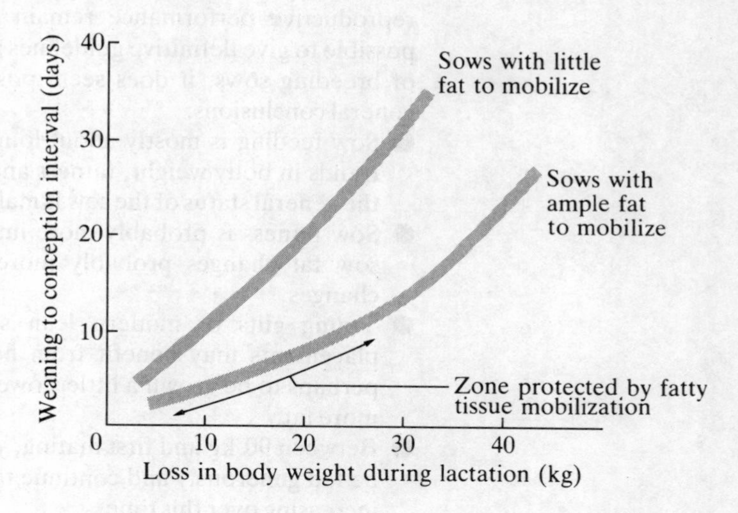

here is that the adverse effects of weight loss upon re-breeding proficiency will be mitigated by the presence of adequate fatty tissue stores.

Problems with sow fertility should not, however, invariably be seen as soluble by nutritional means. Other vital factors are the hormone status of the sow and the efficacy of the boars. Failures in achieving targets for numbers born (11 piglets +) and for farrowings to first service (85 per cent +), may often be resolved by weaning one half of the litter 2–3 days before the other, using a vasectomized teaser boar, using more than one boar at the same oestrus, mating at least twice, maintaining sow health after parturition and ensuring maximum immunity to disease organisms by arranging a substantial challenge early in the life of the prospective breeding sow.

All experiments completed thus far have prevaricated about what the recommended average herd levels of weight, fatness or condition ought to be. It does seem likely that some extremely well managed herds, whose sows are kept in good environments, may well achieve a high level of reproductive success at weights, fatnesses and condition scores rather lower than other herds, whose management and environment may be of a lower standard.

Conclusions

While the particular problems of condition scoring have yet to be resolved, and the relationships between nutritional strategy and reproductive performance remain clouded, and while it is not possible to give definitive guidelines as to the nutrient requirements of breeding sows, it does seem possible to come to some broad general conclusions.

- Sow feeding is mostly about long-term strategy and long-term trends in body weight, fatness and condition. It is essential that the general status of the sow is maintained at a satisfactory level.
- Sow fatness is probably more important than sow weight; and sow fat changes probably more important than sow weight changes.
- Young gilts of modern lean strains destined for herd replacements may benefit from being reared in a special way: perhaps to be grown a little slower and encouraged to be a little more fat.
- Between 90 kg and first mating, young breeding females should be fed generously and continue to grow actively, their fat levels increasing over this time.
- Young breeding females that are tending to be lean should probably be mated no earlier than 110 kg and may benefit from waiting even longer, provided body condition does not fall during this time.
- Excessively high feed levels in pregnancy may reduce embryo survival and decrease lactation appetite.
- Pregnancy feeding should be adequate to allow gains of sow maternal body weight and fatness, in particular to achieve a condition score of around 6 (on a 10-point scale) by the time of farrowing. This will be approximately equal to a P2 fat depth of 14–20 mm and a total body gain from conception to pre-partum of about 40–50 kg. This usually means a daily feed allowance of more than 2 kg.
- Fatter and heavier sows at parturition will have lower appetites, and will lose more fat and more weight during lactation.
- Sows can be fed during lactation to appetite; that is, be presented twice daily with as much food as they will eat.
- Sows can be fed generously (more than 3 kg) or to appetite between weaning and mating.
- Sows that are in good condition and that have not lost excessive amounts of fat during lactation will re-breed more readily than very thin sows.

- Over-fatness will result in both diseconomies and reproductive inefficiencies. Sows of condition score greater than 6 should be restricted. Very generous levels of feeding in pregnancy should be avoided by ensuring the correct general strategy, which should avoid the need for short-term crash-course tactics to fatten sows up or to slim them down.
- Sows will unavoidably lose fat during lactation and should therefore gain fat in pregnancy to compensate. Fat levels should be maintained at a steady average state of about 12–16 mm over the reproductive life. Fat level minima are probably in the region of 15 mm P2 at the end of pregnancy and about 10 mm P2 at the end of lactation. Sows over 20 mm P2 may be considered over-fat.

Definitive rules for feeding breeding females are still elusive. Attempts to define nutritional requirement, and to prescribe blueprints for feeding strategies and tactics, have been disappointing. Sow variability in response to nutrients has much to do with the problem. The understanding of the relationships between nutrition and body fatness, nutrition and lactation, and nutrition and reproduction remains one of the most challenging research areas in agricultural science. Until the challenge is met, guidelines for feeding strategies for breeding females must remain tentative, and the means of achieving set targets must remain discretionary and in the hands of local knowledge. Such discretions may extend deep into the fundamentals of pig science, such as the required increase in feed level to achieve a given improvement in condition score and the required condition score to achieve a given reproductive performance.

Supplementary feeding for the sucking pig

Mammalian milk is rich in energy. Cows' milk is a familiar, but atypical, standard with only 130 g of solids per kilogram and 3.2 MJ/kg. Sow milk, with 200 g of total solids per kilogram and 5.1 MJ/kg is slightly more nutritious than ewe milk, and as such is the richest of all domestic animals with the exception of the rabbit. Piglets start life with 1–2 per cent of fat in their bodies. Daily fatty tissue growth soon greatly outstrips that of protein. In the first days of life, perhaps 30 per cent of the gain is as fat. By 2 weeks of age in both lambs and piglets an even greater proportion of the gain is as fat. In consequence, by 3 weeks of age the live body of milk-fed piglets usually contains 16 per cent or more lipid (Fig. 5.19).

Figure 5.19 Chemical composition of milk-fed piglets. The water content at birth, 5 and 10 kg is 81, 68 and 66% respectively. Ash levels are steady at about 3%.

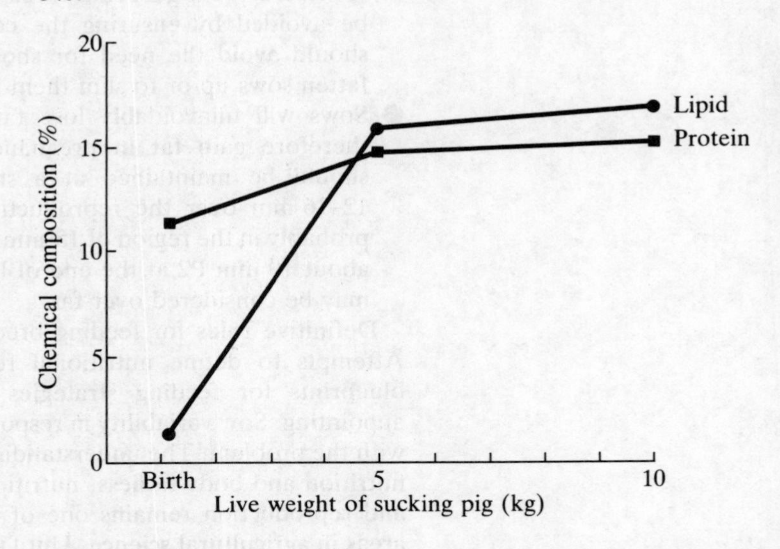

Figure 5.20 Influence of litter size upon milk yield of the sow.

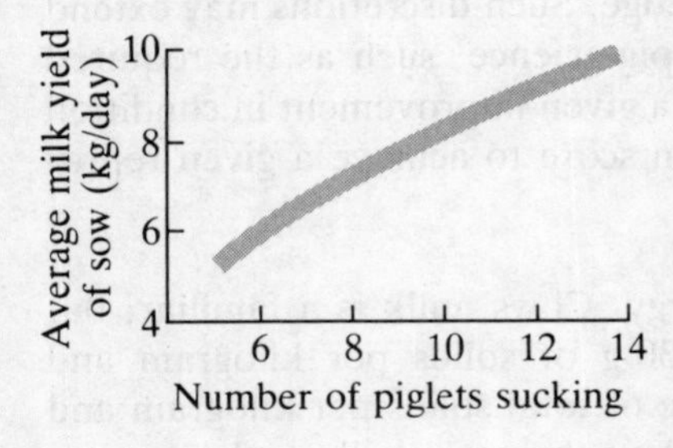

Potential milk yield depends upon the energy intake of the sow and the extent of her body fat stores. Actual milk yield is a further function of the number of piglets sucking, the size and vigour of those piglets, and the stage of lactation (Figs 5.20 and 5.21).

Vigorous sucking pigs in small litters of six or less may readily attain live weights of 10 kg or more at 28 days of age, so it is evident that in the best conditions the supply of milk from the individual mammary gland needs not be limiting to piglet growth up to this age. Usually, however, most sucking pigs from litters of 10 or more weigh only 7 kg at 4 weeks, indicating that there is a shortfall either in total milk supply at the udder or some non-nutritional cause of reduced growth – such as chronic or acute ill health. Given the absence of disease, the pattern of potential pig growth and of sow milk yield diverge at about 3 weeks of age, and it may be assumed that from this point on, supplementary feeding is beneficial, even to pigs in litters small in number. Should the litter be of greater than eight piglets, or the sow be yielding less than potential, then earlier feeding of sucking pigs is warranted, usually from about 14 days of age – but, of course, in cases of milk shortage from very much earlier.

Figure 5.21 *Lactation curves for sows: (a) is appropriate for a large litter sucking a well fed sow.*

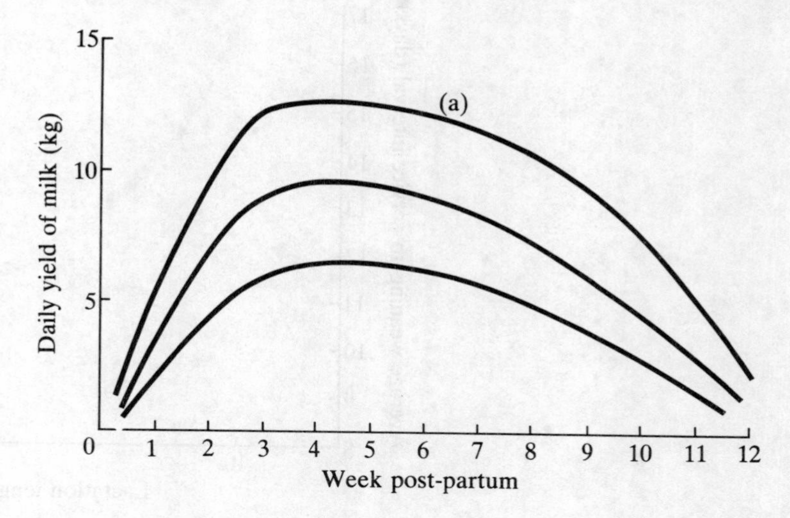

In many studies, piglets sucking sows of good milk yield have been shown to ingest little or no supplementary feed before 18–21 days of age.

The recommended age of piglets at weaning remains a matter of considerable controversy. Most authorities have come to the conclusion that the extra costs of higher standards of housing and management, together with the inverse relationship between length of lactation (7–28 days), and both weaning-to-service interval and numbers born at next litter, result in there being little economic difference in weaning ages between 18 and 28 days (Figs 5.22 and 5.23). In many countries, unavailability of high-quality stockmen, buildings, equipment and diets would indicate 35–42-day weaning as appropriate.

Three-week weaning is currently the accepted European standard, and many pigs are unlikely to have consumed any significant quantity of non-milk food by that time. Normal weaning – indeed, the definition of the word – implies a gradual, not an abrupt, process. Commercial practice therefore places newly weaned pigs in a severely compromised situation.

Sucking pigs have digestive systems different from pigs eating solid food and the changeover at weaning represents a digestive

Figure 5.22 *Influence of lactation length on weaning-to-service interval.*

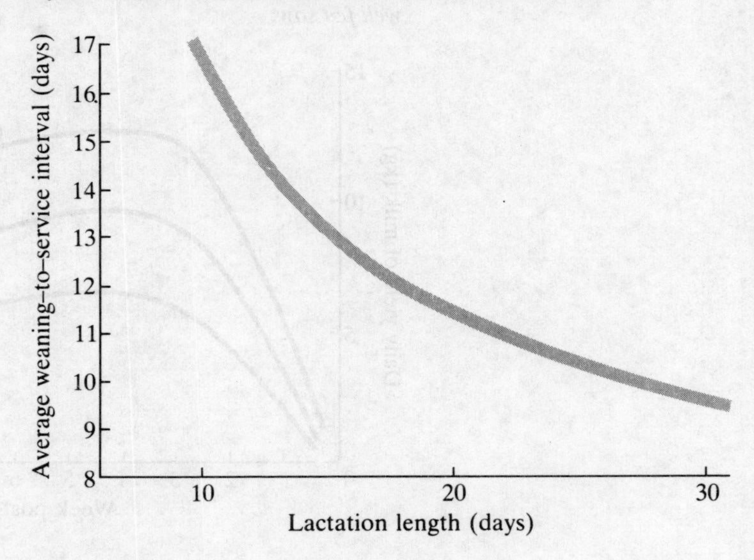

trauma that requires a level of amelioration in proportion to the closeness that weaning is to birth. Natural weaning occurs at around 12 weeks, by which time the pig's behaviour and its digestive system is accustomed to solid food. The young pig will eat solid food as it becomes driven to the nutritional conclusion that the food available at the udder is inadequate to sustain target growth impulsion. When pigs are weaned earlier than, say, 6 weeks of age, post-weaning starvation forces the pig to the realization of the need for solid food. To avoid this counterproductive period, it is necessary to persuade young pigs to eat solids when they have no intuitive drive so to do. Decisions about how much of what sort of food to offer to sucking pigs is particularly complex for those contemplating weaning at 18–24 days. It is almost as if the pig industry had volunteered to opt – for the best of economic reasons – for biological problems whose solutions have yet to be elucidated. Unless the highest levels of animal management skills are employed, the consequences are often either chronic or acute ill health, with all the financial penalties that follow from mortality, morbidity, reduced appetite, lost growth, poor feed conversion efficiency and diminished pig throughput in the post-weaning phase. Weaning at less than 6

Figure 5.23 *Influence of lactation length upon economic efficiency of weaner production.*

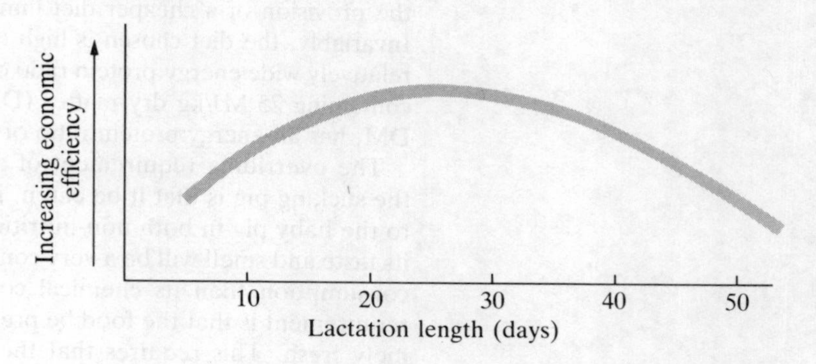

weeks invariably brings about rapid losses of fat from the young pig's body, which occur whilst digestive and behavioural adjustments take place.

Supplementary feeding of sucking piglets is necessary on two counts.

- For litters of reasonable size, to attain the greatest possible growth impulsion.
- For all litters, to accustom piglets to the nature of solid food in order to prepare for early abrupt weaning.

Although it can be taken that mother's milk is likely to be a good match to the needs of the baby pig, a solid diet that mimics milk or is itself high in milk products is not necessarily the first choice to offer when milk is also amply available at the udder. Providing weaning takes place before there is a deficit between the nutrients that the animal requires and the nutrients that are available from the mother, then no solid feed is required before weaning for purposes of proper nutrition. However, it *is* evident that the first choice of a solid diet to offer *after* weaning is likely to attempt to mimic milk or be itself high in milk products. It is only following successful weaning and after the piglet is consuming significant amounts of solid diet that the gradual change in the composition of the diet to cheaper cereal-based ingredients can be begun. In this way the natural (gradual) weaning process could in part be reflected, and the young pigs perhaps totally weaned onto diets largely of vegetable origin by the time they are 8–10 weeks old and 20 kg live weight.

An alternative possibility is to acclimatize the baby pig to nutrients of non-milk origin whilst it is still sucking and thus pave the way for the provision of a cheaper diet immediately upon abrupt weaning. Invariably, the diet chosen is high density and high protein, with a relatively wide energy:protein ratio of about 1:16 DE:CP (sows' milk containing 25 MJ/kg dry matter (DM) and 300 g CP per kilogram DM, has an energy:protein ratio of 1:12).

The overriding requirement of a supplementary solid food for the sucking pig is that it be eaten. It must therefore be interesting to the baby pig in both non-nutritional and nutritional terms, and its taste and smell will be a very considerably greater stimulation to consumption than its chemical composition. Perhaps the prime requirement is that the food be presented frequently and be extremely fresh. This requires that the diet be compounded in small batches and as nearly as possible to the time of presentation to the piglets. Feed should be stored away from pig odours.

When weaning takes place at 6 weeks of age or later, the young pig may be eating sufficient supplementary solid food such that upon weaning it can immediately sustain itself. This is the conventional logic of supplementary creep feeding, but weaning at around 3 weeks makes such a proposition highly unlikely. Whilst it appears self- evident that the provision of supplementary feed to sucking pigs from as early in life as possible is bound to be a good thing, there has recently been raised the proposition that a little may be worse than none at all. The gut lining and digestive system require to adjust to the particular characteristics of vegetable rather than milk proteins. The, as yet unproven, proposition is that, when the young animal has received only a small intake of vegetable protein, the gut is sensitized but not adjusted. This delicate position passes with the consumption of further amounts of vegetable protein and the system stabilizes, ready for survival independently of mother's milk. However, should weaning occur at the very point of sensitization and the gut be challenged by massive doses of vegetable protein at this time, then the consequences can be enteric disorders, loss of appetite, diarrhoea and the invasion of disease organisms. Unfortunately for these propositions, it appears that the total amount of supplementary feed that is required to get the young sucking animal over the sensitization period and into the fully adjusted period is so great that young pigs would be most unlikely to have consumed the requisite amounts by 4 weeks of age.

If, after weaning, baby pigs are to avoid the consequences of starvation, then on the day following removal from the sow they require to be consuming about 300 g of solid food. Converting at 1:1, this would give daily weight gains less than would have been occurring on mother's milk at about 3–4 weeks of age, but adequate to avoid fatty tissue catabolism. Such intakes are rarely, if ever, achieved by sucking pigs of less than 30 days of age, even under the best conditions.

In conclusion, fresh supplementary food based on highly digestible cooked cereals, together with highly digestible dietary fats, and dried milk and whey products, can be of benefit when provided to sucking piglets from about 7 days of age. These diets may also contain sweeteners or palatability enhancers and, of necessity, there will be a level of vegetable protein. The objective of supplementary feeding with such a diet is to accustom the sucking pig to the act of ingesting solid food and thus preparing the animal for the nutritional traumas that are to follow upon abrupt early weaning. The same diet should be continued after weaning, if only for the sake of familiarity. The ingredient composition should therefore be dictated by the post-weaning nutritional requirement.

Quality in baby pig diets is dependent first and foremost upon the quality of the original ingredients. Human-grade standards require to be set for all the ingredients, including added fats and oils, cooked cereals, and milk products.

Chapter 6 Simulation modelling: the prediction of growth response to nutrient supply

Pig production as a biological business

The components of success in pig production are no longer good husbandry and high output alone. Pig production is an entrepreneurial activity, exploiting the financial environment within which production, marketing and trading take place. Managers need to know the consequences of an infinite variety of possible actions and combinations of actions on both the production and the financial effectiveness of the business. Husbandry blueprints and global recommendations are by themselves inadequate because they concentrate the manager's mind on the elements of production rather than on the elements of profit. They encourage a static view of targets for production success, when success is more likely to follow from a flexible approach to production goals such as feed conversion efficiency, growth rate and carcass quality.

The choice of pig meat production strategy must depend closely upon the prevailing circumstances for production, and upon the financial and marketing environment. These circumstances are variable between production units, geographical areas and years. Pig producers are obliged therefore to appraise and adjust their production strategies frequently if profit maximization is to be ensured. For example, pigs can be offered innumerable levels of feed allowance, and a plethora of diet qualities and types; they can be slaughtered over a wide range of weights and the end-product can be valued in a variety of different ways. The market environment for the purchase of inputs and for the sale of outputs interact with the production processes of the pig unit; the financial and production environments are interdependent, and not independent, as may have been thought in the past.

Production managers must decide upon production strategies that will give the greatest profit within the wider context of the business environment. Response prediction is a fundamental requirement if the manager is to be sure of making correct decisions. Before deciding upon an alteration in production strategy, the entrepreneur must have a view of the consequence of such changes. Only if the biological and monetary outcomes of a change in production practice is known can a sensible decision be made as to what optimum production practice might be.

Broadly, responses to changes in management practice are predicted by one of two means. The first is on the basis of historical precedence – previous experience has led to the conclusion that a certain action will result in a certain response. Unfortunately, in pig production, this is not a particularly useful means of response prediction as both production and financial circumstances change so rapidly. A method must therefore be sought that is independent of historical data. This second approach requires an understanding of the causal forces of responses. If the nature of the driving mechanisms for a response are understood, then responses may be predicted. It is this last approach that has led to the building of simulation models. Simulation models represent the individual production unit and allow themselves to be manipulated in a similar way as would occur in real life. If the model is good, then the simulated responses and outcomes will mirror the happenings that are likely to occur in reality. By this means, business managers may reject courses of action that may be seen to be counterproductive, and may select amongst feasible and winning solutions.

Response prediction is most effective when the causes of the responses are relatively few, interactions are absent and the system abides to simple laws. (Given that pig production appears far removed from these premises, it is hardly surprising that blueprint recommendations based on historical precedent have, despite their clear inadequacy, found favour in the past.) Causal forces are fewest, interactions least and operational activities closest to simple laws at the lowest, most fundamental, levels of biology. Consequently, applied experimentation, being at the higher biological levels, is not the best route to the achievement of applied response prediction. Rather, it is better to use data relating to the lower layers of the active elements of the system and to operate below the level of the interactions. Appropriate experimentation to predict responses to food intake is more likely to be a metabolism study than a feeding trial.

The production manager about to move from a blueprint, recommended husbandry practice, philosophy over to one based upon response prediction requires that perceptions of the production process be turned upside down. The blueprint approach begins with a definition of recommended diet composition. This may be found in publications such as *Nutrient Requirements of Pigs*, published by the Agricultural Research Council of the UK in 1981 (CAB, Slough), and the sister National Research Council publications from the USA. Diets are then mixed to the given specification and given to a set allowance level, such as is predetermined to provide pigs of a given target quality (Fig. 6.1). The production manager operates on the assumption that the given feed composition is correct and that if the target of pig quality is successfully met, then the production system has been well controlled and profit will automatically follow. Whilst this may have been the case: (a) when the average pig price was fixed in relation to the cost of pig feed, (b) when meat was in short supply and (c) when the producer's view of marketing ceased at the farm gate, it certainly no longer applies in the present commercial agricultural environment. On the contrary, the pig production manager requires to have first a view of the profit level required to cover his particular level and type of investment (Fig. 6.1). This, the profit, is the only valid production target and criterion of success. To achieve this profit, the feed allowance is manipulated toward achieving a pig quality standard that must be flexible, varying according to pig price and feed price. Lastly, the composition of the feed presented to the pigs requires to be derived as a result of the other parameters impinging upon the production process. The concept of nutrient requirement therefore moves from the domain of the nutritional scientist and into the domain of the economist. Provided that the nutritionist can provide the requisite information to predict the outcome of changes in nutrient specification of the diet, then the actual specification used becomes variable rather than given, a function of the internal workings of the pig unit itself and not a function of any grand national scientific committee.

Pig producers are well aware of complexities caused by interrelationships between the factors impinging upon pig growth. For example, increased feeding levels accelerate growth rate and increase throughput, but also reduce carcass quality. The cost benefit of increasing the feed allowance is a most involved

Figure 6.1 Two contrary views of appropriate end-targets and the means toward those ends.

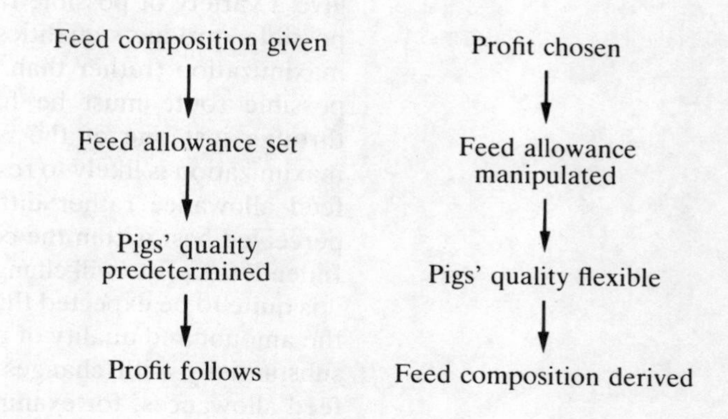

Figure 6.2 Decisions about diet specifications and about carcass quality require all the interrelated factors to be considered simultaneously.

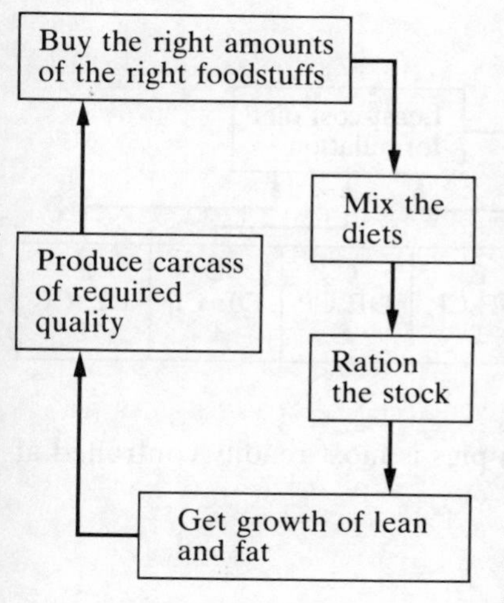

calculation, which cannot be done unless all the facets of the production process are interactively interlinked. Doing only one part of the calculation (such as the relationship between feed intake and fatness) can lead to erroneous conclusions. Because of interactions and knock-on effects, the manager must be simultaneously aware in a quantitatve way of all the factors impinging upon the production process (Fig. 6.2).

The provision of production recommendations on the basis of intermediate goals such as growth rates, feed conversion efficiencies and carcass qualities, is inadequate from the manager's point of view, and is also an unsatisfactory way of conveying the sum of scientific knowledge that is already available. One such intermediate goal is that of nutrient requirement. Diet specifications, agreed by scientific committees, may be presented as targets for diet formulation. One of the earliest uses of the computer was in linear programming to least-cost diet formulation. By allowing comparative pricing of all the nutrients in available foodstuffs, least-cost diet formulation has been the major route to satisfying a given recommended nutrient content in pig diets (Fig. 6.3). The implication here is that if the stated nutritional content of the diet is achieved at least cost, then the ultimate economic goal is achieved. Producers, however, should have only passing interest in the nutrient content of the diet as this is but an intermediate goal and one that should be derived rather than given. A more realistic statement of the situation is given in

Fig. 6.4, which shows that there are a range of possible nutrient specifications and a range of possible feed allowances, which may give a variety of possible rates of meat production at a variety of possible product qualities. To achieve the goal of margin maximization (rather than nutrient content of the diet), the best possible route must be found through the entire system, not through just one of the component parts. The goal of margin maximization is likely to result in a diet nutrient specification and a feed allowance rather different from that that may have been perceived best within the context of a blueprint recommendation. Indeed, the route will change with changing financial fortunes, and it is quite to be expected that both nutrient content of the diet, and the amount and quality of pig meat produced, may require to vary substantially with changes in market environment. Appropriate feed allowances, for example, are highly time dependent.

Figure 6.3 *Least-cost diet formulation to optimal solution of a given nutrient specification (DE, digestible energy; CP, crude protein).*

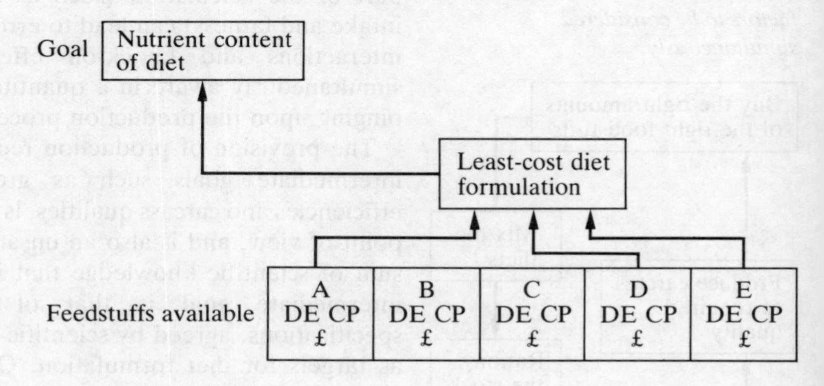

The production of meat from pigs is most readily controlled at four points (Fig. 6.5):

Diet formulation control
Feed intake control
Genetic control
Product quality control

The proportions of individual foodstuffs chosen controls the diet specification, while the feed intake control meters the amount of feed allowed to the pigs. Genetic control governs the type and

Figure 6.4 *Production system analysis to margin maximization by a variety of possible routes.*
(From Whittemore, C. T. (1981) BSAP Occasional Publication, No. 5, *p. 47)*

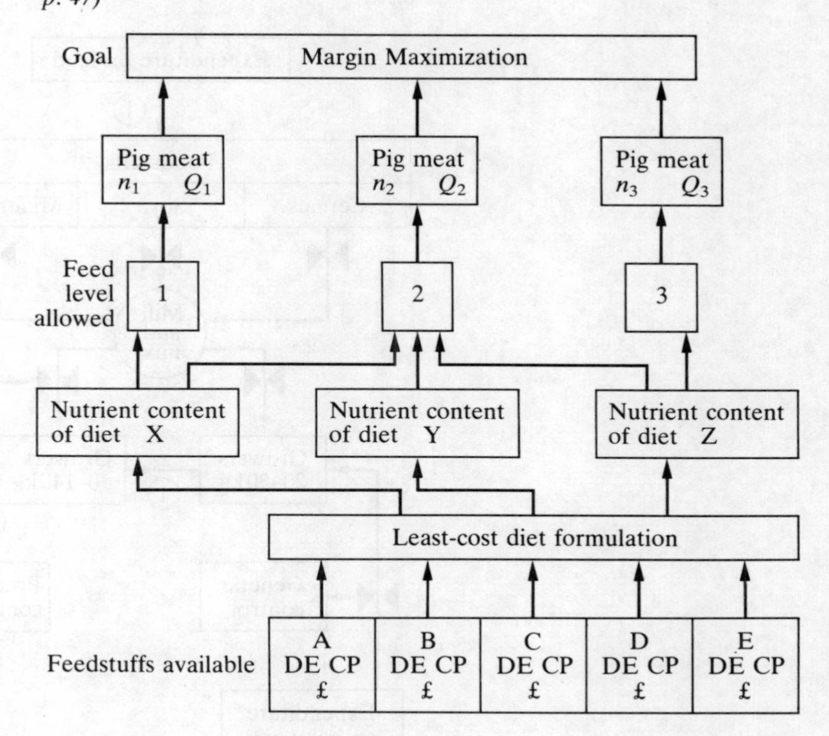

quality of the pig, which will influence the way feed is utilized. Product quality control, exercised at the point of despatch to the meat packer, allows an influence not only upon the fatness chosen as appropriate, but also the slaughter weight, the sex and the sales outlet selected. In order to monitor biological and financial progress, information is needed on expenditures and income, and additonally upon the weight of animals as they grow from 20 kg through to slaughter (Fig. 6.5). Information as to the nutrient specification, feed intake, pig type and product quality will be available through the mechanisms used for control at these points. Given such information, a response prediction model simulating pig growth can be used to guide the manager towards taking the correct production decisions at each of the four control points;

Figure 6.5 *Flow diagram of a grow-out unit, showing essential information collection* Ⓘ *and control* ▶◀ *points.*
(From Whittemore, C. T. (1981) BSAP Occasional Publication, No. 5, *p. 476).*

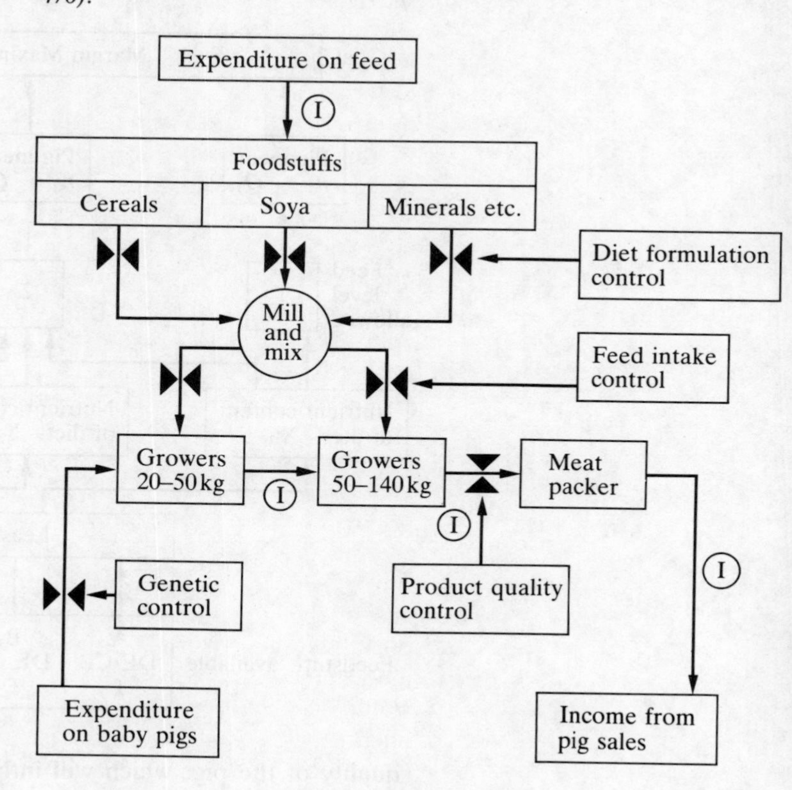

namely, to decide upon the nutrient specification, the feed allowance, the pig type to be used, the weight of the pig at sale, the fatness of the pig at sale and the meat packer selected.

In addition to acting as a decision-making aid, simulation models may also be used for purposes of transferring information and identifying gaps in knowledge. Computer simulation models can contain within them the full range of current scientific awareness about a subject, and bring together separated disciplines and experimental results. Such a program may serve as an information library that does not require either to be read or understood; merely used. A highly sophisticated methodology can by this means be made available to end-users who could not be expected to approach its comprehension.

This facility of short-circuiting the need to understand deep science may also open such models to misuse. The model builder will have had to make choices and to have included information of various degrees of validity. Because models require information at every step in a chain of events, sometimes the modeller must interpolate to fill gaps and the user may be unaware of such weaknesses. Nevertheless, whilst it is important that model builders are honest with model users about the weaknesses and strengths of their products, a model need not necessarily be exactly right in order to be helpful; it is sufficient that a new methodology is some improvement upon that which it supersedes. There is little doubt that, from this point on, the computer revolution will ensure that the vast proportion of information transfer between science and practitioners will be through the medium of simulation models and other such computer packages. If the need to complete information strings is a shortcoming in a simulation model system, it is an advantage inasmuch as gaps in knowledge are effectively identified. Thus researchers may be pointed towards areas of ignorance where further experimentation is required. Sensitivity analysis of the model components will also throw light upon which aspects of the system require to be known accurately and which only approximately. It is sometimes the case that the pursuit of science may encourage individuals to be ever more precise in areas of knowledge where an approximation would do. Alternatively, vital issues may be left to one side by the scientist as being too difficult or not scientifically fashionable. Computer models can assist in better organizing research resources so as to avoid the study of non-problems or the refinement of measurements that no longer need to be refined. Similarly, issues of vital interest and importance are highlighted by a simulation model showing them to be a sensitive variable within the system.

Model-building approaches

Simulation models usually exercise themselves through the medium of computing, and are made up of algorithms comprising mixtures of text statements and mathematical equations. Relationships within models can range from empirical regression derived from feeding trial data (for example, live-weight gain on food intake) to deductive sequences of mathematical descriptions relating to fundamental biochemistry and physiology. Empirical regression tends to be static and inflexible, and strictly best used for finding intermediate points within a data set. There is no reason to believe that a regression relationship will necessarily throw light on the nature of the causes of the responses achieved.

Regression relationships linking inputs to outputs can be of three general types:

(a) The linking of two parameters with high statistical accuracy but using coefficients that are biologically nonsensical.
(b) Or, next, and more satisfactorily, the constants and coefficients within the equation may be reconcilable with biological expectations.
(c) The third type of relationship differs from the other two by linking inputs and outputs through the medium of their causal forces; that is, deduction is employed rather than empiricism.

For the deductive methodology to be effective, intimate knowledge is required of the system. Therefore, although more scientifically acceptable, the deductive approach has of necessity to contain a mixture of hard fact, soft fact, hypothesis and inspired guess-work.

Working models tend to be made up of mixtures of empirical and deductive relationships. But the greater the extent to which the model relies upon the deductive approach, the more likely it is to be truly informative. Deduction – building up input/output relationships from a knowledge of the causal forces – increases the flexibility of the model and allows prediction outside the circumstances within which the information was collected. Such a model can therefore account for a wide range of interactions. The same deductive approach also helps in the understanding of the system, it highlights ignorance, and allows sensitivity analysis and definition of crucial components.

Deductive modellers stand or fall on their ability to form hypotheses about the nature of life, as these will control the algorithms. Growth modelling requires to be pursued at the level of the basic principles. Preconceived notions of growth potentials and the use of empirical analyses of feeding-trial data are likely to lead to less effective simulation. Deductive models employ what is known of the science of nutrition and growth, and the necessary hypotheses reflect the nature of the science that goes into them. The deductive model will thus go further than merely redescribing phenomena previously observed. It avoids simplistic recounting of history and attempts to foresee the likely outcome of future activities. There remain, however, many phenomena that require to be described within a model but for which causal forces are not understood. In this event, relationships can only be approached in an empirical way.

The empirical and deductive approaches may be contrasted with the following example. It is accepted that as pigs grow bigger, they can become fatter. Regression relationships can be drawn up with fatness as a function of live weight. Using this relationship, increasing fatness may be predicted as a consequence of increasing weight. This empirical correlation would lead to the conclusion that slaughter at a heavier weight would invariably result in fatter carcasses. On the other hand, to obtain a deductive function between fatness and live weight requires the causative forces to be identified. One reason why animals may get fatter as they get bigger is because their appetite has increased and the extra food consumption has gone into the production of fat. This, of course, will only happen if the potential rate for lean growth has been reached, therefore there is an interaction with pig type and quality. By considering the causal forces, fatness is seen to be not a direct function of live weight but is flexible, and may be manipulated according to feed intake and genotype. If required, there need be no relationship between fatness and live weight, animals of high genetic potential being able to eat to appetite and remain as equally as thin at 100 kg as they were at 20 kg. The deductive approach presents the production manager with a different set of propositions as to how the pig carcass may be made more lean.

Figure 6.6 The relationship between feed intake and growth: an empirical approach.

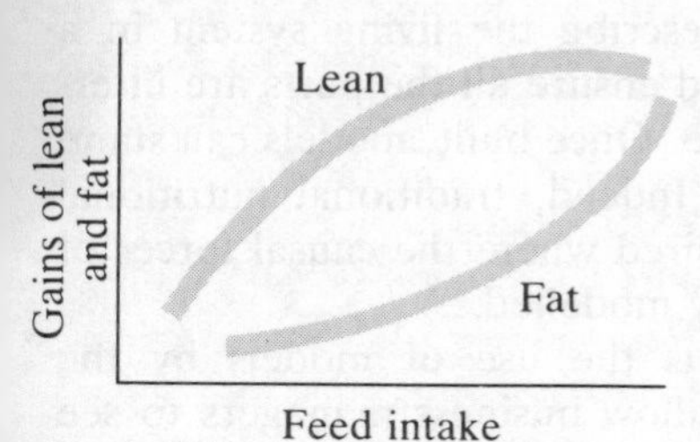

It is evident that the empirical statement of the growth response to feed intake shown in Fig. 6.6 (whilst correctly reflecting a set of experimental data) is quite inadequate for effective simulation of the type required by production managers. The deductive approach is shown in Fig. 6.7 in the form of a flow-diagram describing the system relating feed to growth. This second approach is more complex, but is more soundly based in science and is more likely to simulate real life responses under a wide range of circumstances. It follows that solutions to all operational relationships need to be solved simultaneously. So in seeking a relationship between growth and feed intake, relationships between feed and maintenance, feed and cold, and the metabolic pathways for energy and protein partition must also be understood.

Figure 6.7 The relationship between feed intake and growth: a deductive approach.

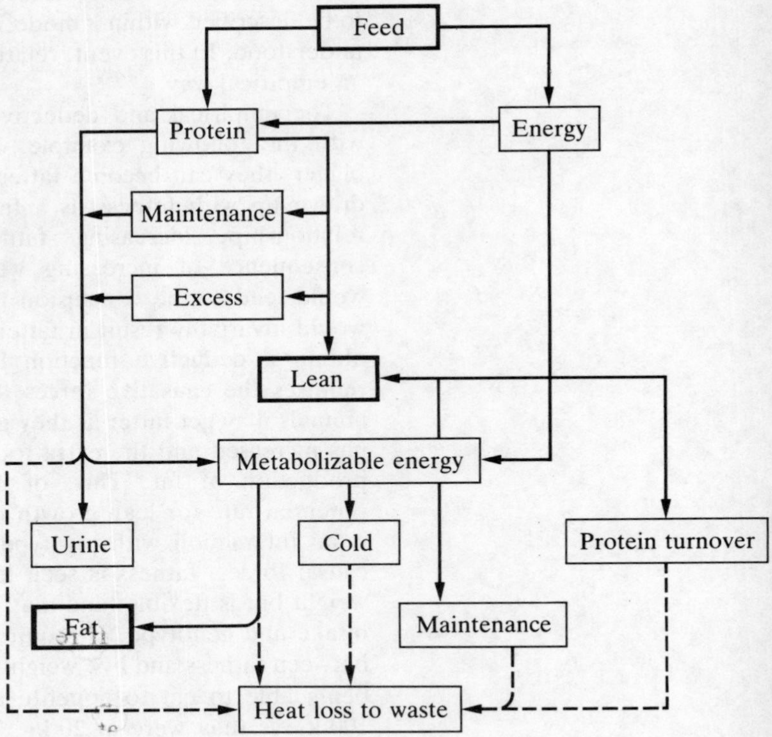

Feed
Protein
Energy
Maintenance
Excess
Lean
Metabolizable energy
Urine
Cold
Protein turnover
Fat
Maintenance
Heat loss to waste

Model building itself is a scientific exercise and biology is better understood through model-building activity. A model forces its builder to complete understanding, otherwise it will not work effectively. The scientist must describe the living system in a consistent and systematic way, and ensure all the parts are interlinked in a way that reflects nature. Once built, models can stand instead of real-life experiments. Indeed, traditional nutritional response trials are no longer required where the causal forces of these responses may be effectively modelled.

But perhaps most important is the use of models by the agricultural industry in order to allow business managers to see ahead. By the use of the model, an entrepreneur may obtain predictions of the consequences of various biological and

Figure 6.8 This schematic diagram of the nutritional affairs in the day of a pig is taken from Scottish Agricultural Colleges Publications, No. 122 *(Pig Farmers Guide:* Publications Nos 109–116; 120–127. *It shows nutrients in terms of gross energy (GE) and protein (CP) in soya and barley being formulated into a diet with 16 MJ of GE and 150 g CP per kilogram. The diet is then given at the rate of 2 kg/day. The digestive system of the pig yields 26.3 MJ DE (digestibility of energy 0.82; realized DE of the diet 13.1 MJ/kg) and 240 kg digestible crude protein (DCP) (digestibility of protein 0.80). DCP used for maintenance, and also that provided in excess, is deaminated (140 g) while the remaining 100 g DCP is deposited as protein in the daily lean tissue growth of the whole pig, which in this case amounts to 0.45 kg. DE used for maintenance (12 MJ/day), together with the DE used to fuel the growth of lean and fat, and to create urine, accounts for 19 of the available 26.3 MJ DE. All this is driven off as waste heat. The remaining 7.3 MJ is found as the energy now locked up in the lean (2.3 MJ) and fatty (5 MJ) tissues of the daily growth. In total, after the growth of bones, and other bits and pieces, are included, this pig will have grown in the day something like 0.650 kg daily live-weight gain.*
(From Whittemore, C. T. (1984). Journal of the Royal Agricultural Society of England 145*: 126–138)*

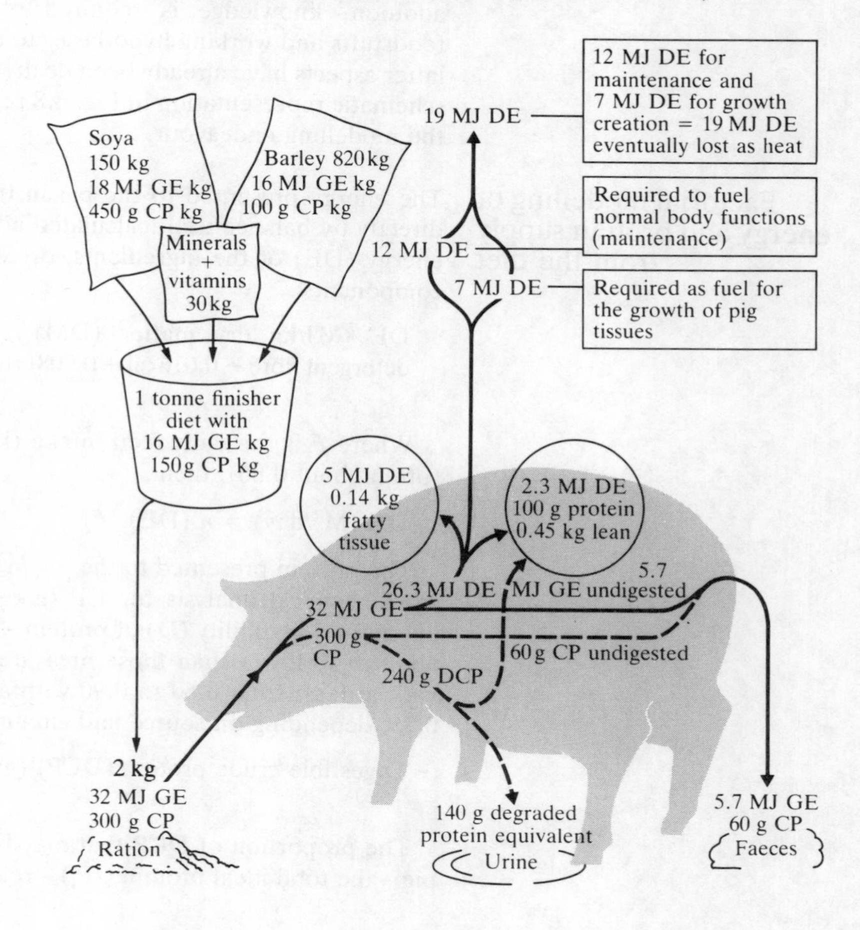

economic activities. By simulating what may or may not happen if this or that management procedure is changed, the manager is given a choice amongst production strategies. Simulation models are uniquely effective at making biological knowledge work toward economic advantage.

The quest for the fundamental elements as described in Fig. 6.7 underlines the modelling approach used at Edinburgh over the last decade. Surprisingly few basic pieces of information are required: such as those that relate to the energy cost of maintenance, the energy cost of keeping warm, the energy cost of protein and fatty tissue deposition, the energy cost of protein deamination, the protein cost of maintenance, and the protein cost of lean gain. In addition, knowledge is required of the nutrient yield from foodstuffs and working hypotheses to describe growth. These two latter aspects have already been dealt with in earlier chapters. The schematic representation in Fig. 6.8 represents a starting point for the modelling endeavour.

Factorial modelling of energy and protein supply from the diet

The energy presented to the pig in the diet may be determined directly by balance trial, calculated additively from the digestible energy (DE) of the ingredients, or estimated from the chemical components:

DE (MJ/kg dry matter (DM)) = 17.5 − 0.015 neutral-detergent fibre + 0.016 oil + 0.008 crude protein (CP) − 0.033 Ash [Eqn 6.1]

Where F is the daily feed intake (kg) of air-dry feed at n DM (often about 0.86), then:

DE (MJ/day) = n (DE) (F) [Eqn 6.2]

The protein presented to the pig in the diet may be determined from chemical analysis for CP (g/kg) as 6.25 × nitrogen. The apparent digestibility (D) of protein – using ileal measurements of about 0.10 lower than those measured by simple balance – may vary at least from 0.60 to 0.90 with a plausible average of about 0.75, depending on source and circumstance:

Digestible crude protein (DCP) (g/day) = D (CP) (F) [Eqn 6.3]

The proportion of DCP that may be taken to be usable by the pig – the total ideal protein (IP_t) – relates to the balance of amino acids and the resultant biological value (V), which may range between 0.6 and 0.8, with a mean in the region of 0.7. The estimate of usable protein requires to be further adjusted downwards to account for the mechanical inefficiencies ($1-v$) of transfer of ideal protein absorbed to ideal protein used, which appear to result in the losses of some 5–15 per cent ($v \simeq 0.9$):

$$IP_t \text{ (g/day)} = D \text{ (CP)} (F) (V) (v) \quad \text{[Eqn 6.4]}$$

DE ingested is used to yield first metabolizable energy (ME) and next net energy for use in various body functions. An accurate estimation of ME requires the distinction of DE of protein origin, which is less valuable than DE from other sources, first because it is directly deposited in lean tissues (rather than used for energy), and secondly because, before protein can yield work energy or substrates, it must be deaminated. Effective ME (ME_t) is therefore the energy available for work and for energy storage in tissues, but excludes energy lost in urine or used for deamination:

$$ME_t = Qd + 23\ Pr + Epf \quad \text{[Eqn 6.5]}$$

where:

$$Epf = DE - (23\ DCP + 0.05\ DE) \quad \text{[Eqn 6.6]}$$

and is a measure of the protein-free digestible energy on the basis of feed and body proteins having 23 MJ energy per kilogram, and 5 per cent of the DE being lost through the creation of volatile fatty acids of reduced usefulness and gases of zero usefulness in the course of hind-gut fermentation of fibrous elements in the feed.

The parameter Qd represents the energy yielded from protein deamination. Assuming urine to carry 7 MJ/kg of protein deaminated and the work energy needed for urea synthesis to be 5 MJ/kg protein deaminated, then:

$$Qd = 23\ Pm - (7\ Pm + 5\ Pm) \quad \text{[Eqn 6.7]}$$

where Pm is the protein deaminated; that is, all the protein that is digested but not retained in tissues in the process of protein retention (Pr):

$$Pm = DCP - Pr \quad \text{[Eqn 6.8]}$$

Factorial modelling of energy use in the body

Metabolizable energy is either retained in protein or fatty tissues, used for the work of creating these tissues (and bone), or used for the work of maintenance (E_m) and cold thermogenesis. All work energy finally leaves the body as heat.

The concept of a maintenance requirement during active growth is difficult and may not relate closely to energy needs determined when there is zero (or negative) tissue retention. Much of the work of maintenance is likely to be associated with the turnover of protein tissues and it is reasonable to base maintenance on protein mass (Pt) rather than live weight (W). While:

$$E_m \text{ (MJ/day)} = 0.719\ W^{0.63} \quad \text{[Eqn 6.9]}$$

was the preferred equation of the ARC Working Party in 1981 (*Nutrient Requirements of Pigs*, published by CAB, Slough), an alternative such as:

$$E_m = 1.85\ Pt^{0.78} \quad \text{[Eqn 6.10]}$$

may better account for animals of high body lean content, such as entire males.

The energy cost of protein retention (E_{Pr}) has two components, the energy retained in protein (23 Pr) and the work energy used to drive the anabolic processes of protein growth. There is debate as to the latter, which certainly exceeds the price to be paid for simply joining amino acids into increments of new protein. Sticking the components of pig protein together is thought to cost about 5 MJ/kg of protein formed (estimates range from 3.5–8); but the issue turns on the proposition that much more protein is turned over than is newly incremented. In young animals of 20 kg liveweight it is possible that the ratio of new protein accreted to total protein turnover is about 1:5, while in animals of 100 kg the ratio is probably nearer 1:8. Much more energy is therefore needed to break down and rebuild the existing body mass, each kilogram of protein turned over using up a quota of energy. Where Px is the total mass of protein turned over in one day:

$$E_{Pr} = 5\ Px + 23\ Pr \quad \text{[Eqn 6.11]}$$

The remaining element – still awaited – is an estimate for the daily rate of body protein turnover. For the time being, this has to be surmised. It may be assumed that there is a minimal value, probably in the region of 5 per cent of the total body protein mass (Pt).

$$Px_{\text{min.}} = 0.05\ Pt \quad \text{[Eqn 6.12]}$$

It may further be assumed that turnover is some function of the degree of maturity, probably:

$$Px = Pr/(0.23\ (\hat{Pt} - Pt)/\hat{Pt}) \qquad \text{[Eqn 6.13]}$$

where $\hat{Pt}$ is the mature total protein mass. It remains contentious as to whether Px is also related in some proportional way to Pr such that higher daily rates of protein retention are associated with higher daily rates of protein turnover.

Pt and $\hat{Pt}$ require to be assigned some numerical values. Pt is accumulated within the model as the pig grows and is therefore self-generated (as a rule of thumb, Pt usually approximates to 0.16 W). $\hat{Pt}$ is remarkably difficult to quantify, as few experiments have taken pigs through to their mature size and estimated protein mass. What evidence there is suggests values between 30 and 40 kg, the lower relating to females and the higher to entire males. As the daily rate of protein growth is strongly related to mature size, it may be allowable to calculate $\hat{Pt}$ from knowledge of the maximum rate of protein deposition ($\hat{Pr}$) associated with the particular sex and genotype concerned:

$$\hat{Pt} = 300\ \hat{Pr} \qquad \text{[Eqn 6.14]}$$

Fortunately, the turnover of fatty tissues is of less significance than that of protein. Many estimates of the energy costs of fat growth (E_{Lr}) are in close agreement that the efficiency of use of ME for this function is slightly above 70 per cent, with a work energy cost of some 14 MJ per kilogram of fat formed. This cost will be a little lower if a high proportion of the body lipid is formed directly from dietary lipid rather than indirectly from carbohydrate energy components of the diet. Taking the energy value of fat to be 39 MJ/kg:

$$E_{Lr} = 14\ Lr + 39\ Lr = 53\ Lr \qquad \text{[Eqn 6.15]}$$

The heat thrown off from the animal body in the course of energy metabolism goes most of the way towards combating the exigencies of a cold environment. Indeed, there is a necessity for the heat to be carried away from the body if heat stress is to be avoided. Often, however, the environmental temperature may be lower than is required for thermoneutrality, or there may be an excessive rate of heat removal from the pig's body such as might be occasioned by an uninsulated floor or a high ventilation rate. At temperatures (T, °C) below the critical temperature (Tc), energy is required for cold thermogenesis and is diverted away from growth – first from fat growth, and then from both fat and lean growth – in order to satisfy this need. In consequence, pigs grow more slowly in the cold than in the warm.

The dependence of critical temperature upon the heat output (H) from the body can be broadly expressed for pigs of 10 kg or more as:

$$Tc = 27 - 0.6\,H \qquad \text{[Eqn 6.16]}$$

where H is, of course, positively and closely related to the live weight, the feed intake and the growth rate of the pig. The value of H is generated from within the model as the sum of the work energy used for maintenance, and for the growth of protein and fatty tissues; or by difference.

$$H = \mathrm{ME_t} - (23\,Pr + 39\,Lr) + 5\,Pm \qquad \text{[Eqn 6.17]}$$

(That cold thermogenesis itself adds to the same parameter of H, leads to the need for iterative programming.)

$T - Tc$ calculates the temperature deficit to be made good by cold thermogenesis. The energy cost of cold thermogenesis (E_{H^1}) for each degree of cold may be estimated as:

$$E_{H^1} = 0.012\,W^{0.75}\,(Tc - T) \qquad \text{[Eqn 6.18]}$$

Taking proper account of the influence of the quality of the environment upon effective temperatures has proved difficult, the best attempt thus far coming from Dr James Bruce and his colleagues at the Scottish Farm Buildings Investigation Unit at Aberdeen. These workers have quantified, through deep mathematics, the consequence of the rate of air flow over the body and of the rate of heat withdrawal from the body through various floor surfaces of different heat-removal capabilities. The more pragmatic approach may be to simply offer a score for environmental quality along the lines of Table 6.1. The effective environmental temperature is now estimated as:

$$T = T\,(Ve)\,(Vl) \qquad \text{[Eqn 6.19]}$$

and this is the value substituted into Equation 6.18.

Table 6.1 Scores for Ve and Vl for use in calculating the effective environmental temperature (T) in the equation T = T (Ve) (Vl)

Rate of air movement and degree of insulation	*Ve*
Insulated, not draughty	1.0
Not insulated, not draughty	0.9
Insulated, slightly draughty	0.8
Insulated, draughty	0.7
Not insulated, draughty	0.6
Floor type in lying area	*Vl*
Deep straw bed	1.4
Shallow straw bed	1.2
No bedding on insulated floor	1.0
Slatted floor with no draughts	1.0
No bedding on uninsulated floor	0.9
Slatted floor with draughts under	0.8
No bedding on wet, uninsulated floor	0.7

In summary, ME_t (which is already adjusted for deamination losses) is partitioned into maintenance, protein retention, fat retention and cold thermogenesis thus:

$$ME_t = E_m + E_{Pr} + E_{Lr} + E_{H'} \quad \text{[Eqn 6.20]}$$

Factorial modelling of protein use in the body

The total ideal protein absorbed and passed into the body (Equation 6.4) is partitioned into protein required for maintenance (IP_m) and for protein retention (IP_{Pr})

$$IP_m + IP_{Pr} = IP_t \quad \text{[Eqn 6.21]}$$

Maintenance protein requirements result primarily from the inefficiences of protein turnover and comprise the rate of urinary excretion (intestinal losses already being accounted for within the estimate of D):

$$IP_m = 0.004\ Pt \quad \text{[Eqn 6.22]}$$

Given that the estimate of IP_t has included the downward adjustments consequent upon differences between ileal and conventional digestibility, and upon mechanical inefficiency (v), ideal protein not used for maintenance will be fully available for protein retention. So:

$$IP_{Pr} = \mathrm{IP_t} - IP_m = Pr \qquad \text{[Eqn 6.23]}$$

This relationship will hold until $Pr = P\hat{r}$ and the current rate of protein retention reaches the maximum potential rate. Above this point, increasing supplies of ideal protein will be in excess (IP_{xs}) and simply add to the pool of protein destined for deamination (Pm).

Daily protein retention responds linearly to increasing nutrient supply until the plateau is reached. This plateau appears to be related to live weight in a slightly (but unimportantly) quadratic way, as discussed in Chapter 2. Genetic selection for lean tissue growth rate has ensured a continuing rise in estimates for $P\hat{r}$. Current likely values for pigs of 20–120 kg live weight are given in Table 6.2.

Table 6.2 Values for the maximum rate of protein deposition ($P\hat{r}$) in White breeds and strains of pig used for the production of meat (kg)*

	Entire males	Females	Castrates
Grandparent breeding stocks	0.18	0.16	0.14
Improved hybrids	0.14	0.12	0.11
Commercial crosses	0.13	0.11	0.10
Utility strains	0.12	0.10	0.09

* These values are appropriate for Large White and Scandinavian Landrace breeds, and for their crosses. Values for the blockier breeds and for American breeds are likely to be lower.

Factorial modelling of the proportion of fat in live gain

Equation 6.23, showing protein retention (Pr) to equal the amount of ideal protein available for it up to the maximum of $P\hat{r}$ will only hold conditional upon an adequate energy supply for the work of protein deposition (E_{Pr}). It may be assumed that the priority for $\mathrm{ME_t}$ is in the order E_m, E_{H^1}, E_{Pr}, E_{Lr}. This being the case, the daily rate of fat deposition (Lr) can be calculated as:

$$Lr = (\mathrm{ME_t} - (E_m + E_{Pr} + E_{H^1}))/53 \qquad \text{[Eqn 6.24]}$$

While satisfactorily resolving the position when $Pr = P\hat{r}$, Equation 6.24 is inadequate with respect to growth in which Pr is less than $P\hat{r}$, as it implies that normal growth may proceed in the form of protein-containing tissues alone, whereas this is not the case. During nutrient limited growth, even although Pr is not maximized, there is always some fat laid down in the course of physiologically normal gains. The obligatory presence of some fat in live growth will modify the slope of the regression of protein retention on nutrient supply as already descsribed in Fig. 2.16. If the model is to work satisfactorily, a view must be provided of the minimum level of fat that may be expected in nutrient limited gain when $Pr < P\hat{r}$. This is probably best presented in the form of a minimum ratio with protein: $(Lr{:}Pr)_{min.}$. Just as estimates of Pr are increasing in response to genetic selection, then so estimates of $Lr{:}Pr$ are decreasing. Current best estimates are offered in Table 6.3. The $Lr{:}Pr$ ratio is only active while $Pr < P\hat{r}$; as soon as the plateau for the maximum rate of protein deposition is reached then, naturally, the rate of Lr will increase and the minimum ratio will be exceeded.

Table 6.3 Values for the minimum ratio of fat gain to protein gain* $(Lr{:}Pr)_{min}$. *in White breeds and strains of pigs used for the production of meat*

	Entire males	Females	Castrates
Grandparent breeding stocks	0.4	0.6	0.7
Improved hybrids	0.5	0.7	0.8
Commercial crosses	0.7	0.9	1.0
Utility strains	0.9	1.1	1.2

* These values are appropriate for Large White and Scandinavian Landrace breeds, and for their crosses. Values for the blockier breeds and for American breeds are likely to be higher.

Summary of the factorial model for nutrient supply

The model thus far has received a given quantity of ingested feed of given chemical component specification (oil, fibre, protein, ash and important limiting amino acids), simulated digestion and absorption, yielded available energy and quantums of ideal protein, undertaken the work of maintenance, cold thermogenesis and metabolism, and created new growth of chemical fat and chemical protein. These processes can be traced through Equations 6.1–6.24 as presented, while Fig. 6.9 gives a flow-diagram guide to the system analysis protocol.

Figure 6.9 Flow diagram simulating nutrient utilization by growing pigs and allowing prediction of response. Equation numbers are given in brackets.

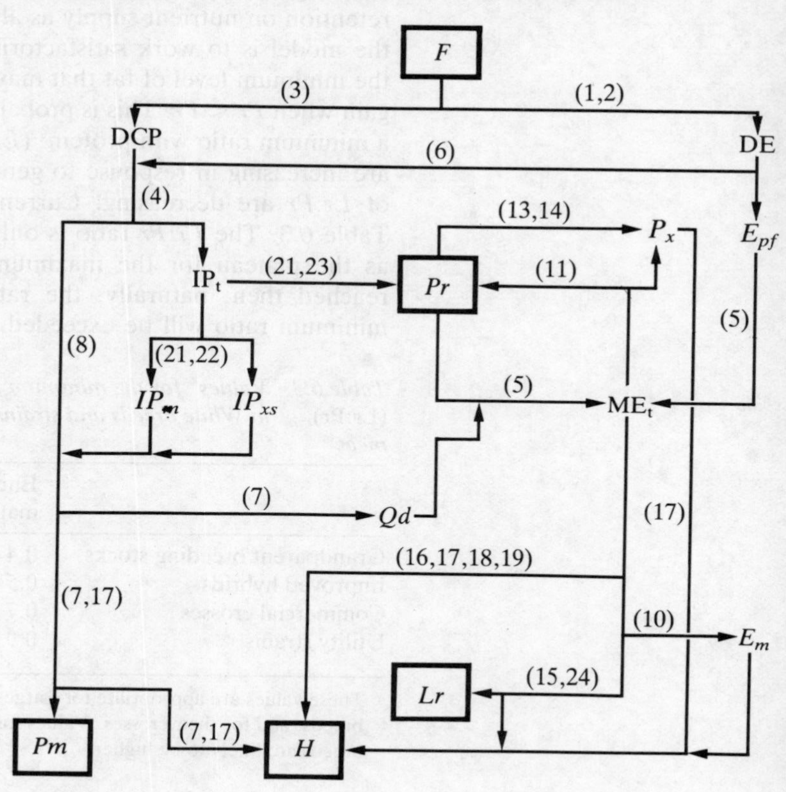

Factorial modelling of the physical components of live growth

At the point of weaning, pigs may contain approximately equal proportions (0.16) of fat and protein. In the post-weaning phase some fat is invariably lost, to be replaced mostly by water. Most pigs of 10 kg, having been weaned at 5 kg will contain about 0.16 protein, and between 0.08 and 0.12 fat, the remainder being water (0.71) and ash (0.03). In visual terms a well rounded weaner will usually contain 0.12–0.16 fat, while a thin one will contain 0.06–0.09 fat. Protein and fat gains (*Pr* and *Lr*), as modelled using Equations 6.1–6.24, need to be accumulated on the base of chemical composition of the pig at the start of the simulation exercise – say at 20 kg. Where *Pt*, *Lt* and *At* represent the total mass of protein, fat and ash at the start of the simulation, then:

$$Pt = 0.16\ W \quad \text{[Eqn 6.25]}$$
$$Lt = 0.07 \text{ to } 0.15\ W \quad \text{[Eqn 6.26]}$$
$$At = 0.03\ W \quad \text{[Eqn 6.27]}$$

Chemical mass can be accumulated during the course of the simulation through to slaughter in the manner: $Pt = Pt + Pr$; $Lt = Lt + Lr$; $At = At + 0.21\ Pr$. The remaining chemical component of the body is water. There is ample evidence showing that as protein mass increases then the amount of water associated with protein in the lean mass reduces. The Polish workers, Kielanowski and Kotarbinska (Kotarbinska, M. 1969. Badania nad Przemiana Energii u Rosnacych swin. Inst Zootech, Krakow, Wydawn. Wlasne no 238), have probably elucidated the most effective description of total body water (Yt):

$$Yt = 4.9\ Pt^{0.855} \quad \text{[Eqn 6.28]}$$

The whole body, empty of gut contents (W_e) is therefore:

$$W_e = Pt + Lt + At + Yt \quad \text{[Eqn 6.29]}$$

Gut contents (ingesta, digesta and large intestinal waste) themselves usually comprise about 5 per cent of the live weight (W), so:

$$W = 1.05\ W_e \quad \text{[Eqn 6.30]}$$

and the total live body weight is thus generated.

The content of dietary fibre, in particular, will influence the estimate of gut content: the more fibre, the greater both the bulk of digesta and the water retained in the gut. Such refinements may be readily made as derivations from the chemical composition of the initially ingested feed. With feeds of lower bulk density, the consequences upon live gain and carcass yield can be significant, a high nutrient density diet often resulting in 1–2 percentage units more carcass yield on the basis of a lesser gut fill alone.

In addition to carcass yield, or killing-out percentage, the important components of carcass quality are the amounts (kg) of carcass dissected lean (Md), carcass dissected bone (Bd), carcass dissected fat (Fd), and the P2 fat depth (mm). Relationships between these characters and Pt and Lt are not well researched, and are certainly dependent upon sex, strain and breed. Dissections and chemical analyses from Edinburgh suggest, as approximations:

$$Md = 2.3\,Pt \quad \text{[Eqn 6.31]}$$

or, an alternative:

$$Md = 2.73\,Pt(Pt/Pt)^{0.076} \quad \text{[Eqn 6.32]}$$

and:

$$Bd = 0.5\,Pt \quad \text{[Eqn 6.33]}$$

or, an alternative:

$$Bd = 2.5\,At \quad \text{[Eqn 6.34]}$$

and:

$$Fd = 0.98\,Lt - 0.14 \quad \text{[Eqn 6.35]}$$

or, an alternative:

$$Fd = 0.89\,Lt \quad \text{[Eqn 6.36]}$$

The depth of fat at the P2 site is related to *Fd* in the alternative forms:

$$\text{P2} = 2.54\,\text{Fd}^{0.68} \quad \text{[Eqn 6.37]}$$

or:

$$\text{P2} = 0.91\,Fd + 0.5 \quad \text{[Eqn 6.38]}$$

For any given value of *Fd*, the blocky strains of pig will usually carry about 15–20 per cent more fat depth at the P2 site. This may be accounted for by adjusting the multiplier accordingly. Other corrections to deal with differences in pig shape and type concern a positive adjustment to the yield of carcass lean, a negative adjustment to the yield of carcass bone, and an improvement in killing-out percentage, all in accord with the precepts given in Chapter 3.

The weight (kg) of the carcass (W_c) yielded relates to both live weight and fatness, killing-out percentage (KO) being higher at higher values of *W* and P2:

$$\text{KO} = 66 + 0.09\,W + 0.12\,\text{P2} \quad \text{[Eqn 6.39]}$$

$$W_c = \text{KO}\,(W)/100 \quad \text{[Eqn 6.40]}$$

The proportions (g/kg) of carcass lean, carcass subcutaneous fat and total carcass dissectible fat are best estimated from P2 using the equations derived by Diestre and Kempster of the Meat and Livestock Commission (Diestre, A. and Kempster, A. J. (1985) *Animal Production* **41**: 383):

Carcass lean (g/kg) = 600 + 0.45 W_c − 7.9 P2 [Eqn 6.41]
Carcass subcutaneous fat (g/kg) = 61 + 7.3 P2 [Eqn 6.42]
Carcass total dissectible fat (g/kg) = 88 + 0.19 W_c + 9.4 P2 [Eqn 6.43]

These may be compared with corresponding empirical relationships derived from Edinburgh dissections:

Carcass lean (g/kg) = 610 − 5.2 P2 [Eqn 6.44]
Carcass total dissectible fat (kg) = 0.19 W + 0.78 P2 − 9.2 [Eqn 6.45]

Factorial modelling of food supply

A statement for the feed intake (F) is a simple matter of programming into the model the given feed allowance scale. This is becoming increasingly straightforward with the advent of computer-controlled diet mixing and wet-feed delivery systems. Most feeding scales increment the allowance on a weekly basis, the size of the increment diminishing as slaughter weight is approached if feed requires to be restricted to ensure adequate grading standards and low backfat depth. Weekly increments usually range between 0.1 and 0.25 kg per pig.

In many cases young pigs are given as much as they will eat between 20 and 40 or 50 kg. Estimating this quantity may present difficulty if *ad libitum* dry feed hoppers are used unless records of feed usage are kept. Under many *ad libitum* feeding regimes, the way in which feed is provided in the hopper ensures that maximum feed intake is not achieved, and thus *ad libitum* feed intakes can vary from restricted to generous, and be highly dependent upon individual farms and management. The amount of food consumed *ad libitum* is usually expressed as a function of metabolic body weight. It is generally recognized that the upper limit to feed intake approaches:

$$F(\text{kg}) = 0.14\ W^{0.75} \quad \text{[Eqn 6.46]}$$

but this clearly can be highly influenced by palatability and bulk density. A rule of thumb is that growing pigs will normally consume about 5 per cent of their live weight daily, but such expectations may be misused (Chapter 1). Pigs allowed free access to feed under commercial conditions may be expected to consume daily 0.12 $W^{0.75}$ when management is good, but only 0.10 $W^{0.75}$ when poor. At Edinburgh (Fig. 2.19) maximum feed intake is quite well expressed as:

$$F_{(max.)} = 0.045W + 0.4 \quad \text{[Eqn 6.47]}$$

with an upper limit (plateau) of 4 kg reached when $W = 80$.

All feeding systems are associated with some degree of feed wastage, and this represents a difference between nutrients offered and nutrients ingested. In wet feeding systems wastage is often about 2.5 per cent, while in some dry feeding systems it can readily rise to above 5 per cent.

Temperature influences feed intake to a considerable degree. Indeed, within the constraints of gut capacity, it is difficult to conceive of controllers upon feed intake other than:

1. The drive of the animal to use the ingested nutrients to good purposes, such as lean tissue growth, fat growth and cold thermogenesis.
2. The ability of the animal to dissipate into the environment the heat created by the metabolism of the ingested nutrients.

It follows therefore that appetite will increase with increasing growth potential and decreasing environmental temperature. Attempts to quantify the negative effects of a high environmental temperature on appetite have been disappointing. It may be surmised that the percentage rate of reduction of expected feed intake is related to $(T - Tc)$. Feed consumed is probably reduced by 2–5 per cent for each degree Celsius that the environmental temperature is above the critical temperature, as estimated by Equation 6.16.

The Edinburgh model pig

The Edinburgh model pig has at its core algorithms and hypotheses similar in nature to those described in this and previous chapters. The content is, however, more sophisticated, detailed and elaborated. The surrounding structures of the model, required to allow the core to operate in a realistic environment, are of considerable scale and complexity; but this is necessary to allow model inputs and outputs to be framed in terms of normally expected parameters of pig production management practice.

The ancestry of the Edinburgh model pig lies with theoretical deliberations at Edinburgh that were initiated in the early 1970s (Whittemore, C. T. and Fawcett, R. H. (1974) *Animal Production* **19**: 221) and that are still continuing vigorously. Since that time, the model pig has been used by research workers, teachers and students, and widely by the industry (feed compounders, breeding companies, advisory services, consultants, veterinarians, farm businesses, pig growers and finishers, and pig procurers and slaughterers). In 1984 a micro version was made available in the form of a floppy disk to run on IBM machines and this version incorporates substantial quantities of new information, in addition to allowing access by the growing community of microcomputer users and creating a program of particular relevance to the contemptorary pig industry.

By simulating the production process, the model allows the manager to make short-term (tactical) and long-term (strategic) decisions on the basis of improved information – and with proportionately reduced risk. The financial and production consequences of alterations in management practices are both equally considered; but the bottom line of the model is financial rather than biological.

The model does not come to a single 'optimum' solution. Rather, it lays out the options that are open to a decision-maker and guides as to the likely outcome of decisions. In essence, the model answers 'What if' questions, such as:

- What if a better, but more expensive, feed is used?
- What if more (or less) food is offered to the pigs?
- What if a different type or quality of stock is involved?
- What should the appropriate tactics be in response to a change in weaner price or feed price?
- What are the effects of different environments upon performance?
- What if the average price of pig meat changes, or the grading penalties alter?
- What if the fixed and variable cost structure changes in relation to overall costs?
- What if the contract scheme for pig sales from the farm is revised?
- How should a manager choose from amongst a range of available contract schemes, or a range of available feed manufacturers?
- How important is rate of growth as compared to carcass quality (grading profile)?
- What would be the effect upon the industry of a change in pricing structure for raw materials and/or products?
- And so on.

Although the model does indeed simulate the growth responses to nutritional supply within the context of the farm and marketing environment, its main purpose is not to reproduce the real-life situation with complete accuracy. This would only be reasonable, in any event, if happenings within the pig world itself accurately repeated themselves. There is, of course, always biological variation in the system and apparently identical circumstances will, in different places or at different times, result in different responses.

The main purpose of a model is rather to show the direction of the response, the rate of the movement of the response and the sensitivity of the production system to a tactical or strategic change (see Fig. 6.10). For example, altering from *ad libitum* to restricted feeding might be shown by a model to reduce growth rate from 900 to 750 g/day and to decrease backfat thickness from 15 to 14 mm. What is important is not so much that the simulated growth rate was identical to that in real life at 900 g before the change, but rather that growth rate was reduced by 17 per cent, whilst fat was reduced by only 7 per cent.

Figure 6.10 *The model should initiate from a position within range of the existing real-life circumstances (i.e. within the broken-line boundary). An exact overlay is rarely achieved and not required; it is only necessary for the model to be started from a similar point to that perceived in real life. If, however, the starting point for the model differs widely from that in real life, then the magnitude and direction of the response can be wrongly predicted.*

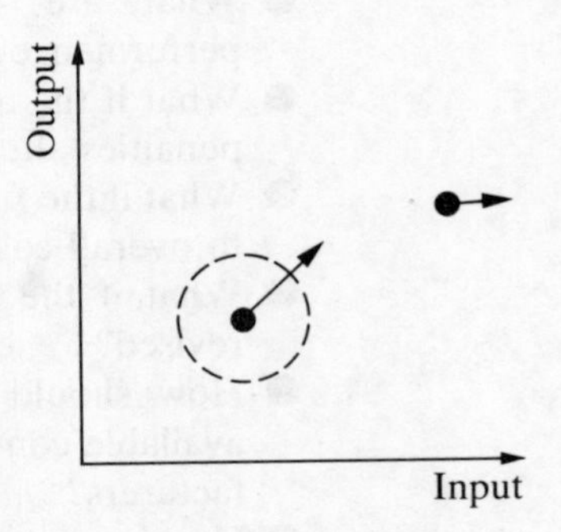

Nonetheless, it is helpful if a model can be tuned in as closely as possible to the current production situation. This is because responses *do* change according to their position on the scale (Fig. 6.10). Growth rate is more sensitive to change in feeding regimes at lower growth rates than at higher growth rates; whilst 1 mm of backfat depth would be considerably more important if the pigs were near the upper limit of the grading requirement.

In general, therefore, the results from a simulation model should reflect, reasonably closely, the real-life situation that is being simulated. However, exact replication should not be looked for, nor is exactitutde a necessary prerequisite for the model to fulfil its purpose.

The instability of pig próduction calls for the model to be referred to when: performance (financial or pig) is less than required; when the manager is contemplating the benefits of a change in production system (feed allowance, feed quality, feed supplier, pig type, contract scheme or whatever); or when the circumstances of production change such that new stratagems are likely to be needed to optimize production under the new conditions – for example, the price received per kilogram dead weight may have altered, or the feed price, or the weaner costs and so on.

Because of the interactions, it is always necessary to do a number of runs to get a feel for the situation and to provide the required navigational aids to decision-making. The type of procedure that might occur would be as follows.

- Returns have fallen upon the imposition of a more severe carcass grade penalty against fatness. Now only 40 per cent of pigs attain top grade. What should the target grading profile be for maximum profit?
- A number of runs are needed to tune in the model as far as possible. If there are large discrepancies between achieved performance and predicted performance, then the model should be used to help diagnose the reasons. Are the pigs failing to attain the top grade through failure to meet other carcass criteria, such as dead-weight limits? If the pigs are so fat, is the feed properly described in terms of its energy and protein content, and the level of feed allowance?
- Having tuned in the model pig, a series of runs would sequentially answer these sorts of questions.
  1. Would grading improve if the pigs were fed less? How much less should be given, and what are the implications for throughput and annual profitability?
  2. If feed were reduced, should it be reduced by the same amount for females and castrates?
  3. Would a reduction in slaughter weight solve the problem?
  4. What would happen if feed costs were reduced by the purchase of a cheaper foodstuff?
  5. Would grading be improved by the use of a more expensive, higher-protein feed compound?

6. Would profitability be improved if better stock were brought into the herd?
7. What would be the benefits of going for a contract that accepted entire males? How should they now be fed?
8. If the overall rate of growth were increased, how many extra weaners would need to be contracted for the unit in the year?

The appropriate line of investigation becomes evident to the manager as he runs the simulation program. This will depend upon the relative sensitivities, which change as production circumstances alter. Each series of runs gives unique answers for any particular year, pricing structure and farm. Therefore, the model pig must be run anew for each new query. Historical information from previous occasions or from previous circumstances is likely merely to mislead.

Experience has shown that models can be used effectively for diagnosis as well as prediction. Where there are significant and irreconcilable differences between responses predicted by a model and occurrences in real life, it is worth querying the cause. Perhaps the description of the real world provided to the computer model was inaccurate in some respects, or perhaps there was some hidden problem. Examples would be: wrong assumptions about the environment, a high prevailing level of disease, feed quality or feed level different from assumed, failure to take account of feed wastage, or misinterpretation of the reason for carcass out-grades. By running the model within the circumstances that are thought to be operating on the unit, a guide is obtained as to expected performance. This performance level can be set up as a target for the unit. Comparison of performance achieved with target set allows the manager to identify points of weakness in his existing management.

Inputs and variables

The model needs information – about the feed, the pig, the farm and the market environment – if it is to simulate and predict. This information has to be as accurate as possible and the *quid pro quo* is that the information be as simple as possible to collect. The more complex the workings of the model, the more likely it is to be able to operate on simple inputs and to generate simple answers.

Every input constitutes also a variable and therefore a manipulable control point. For example, input information is needed about CP content of the diet and the quality of pig used: both these inputs are controllable variables that can be altered by the manager in the course of his pursuit of maximum profit. The following input variables are used in the Edinburgh model. A variety of sequential values may be used for each input over the various phases of growth from start to finish.

Feeding regimes: Feeding method; daily feed allowances (if not *ad libitum*); feed wastage.

Feeds definitions: Fibre; protein; protein digestibility; oil (or DE); protein value (*V*); cost.

Fixed and variable costs: Annual fixed costs for the unit; variable costs per pig.

Pig types and strains: Number of weaners available to the unit per year; weight of weaner at start; cost of weaner at start; sexes to be used; genetic quality of pigs (breed and extent of breed improvement); blockiness (proportion of blood from meat-type sire-lines).

Environments: Ambient night temperature; degree of house insulation and draughtiness; type of flooring; number of pens; number of pigs in each pen; pen days empty between batches.

Market descriptions: Minimum and maximum dead weight requirements; method of payment (flat rate, yield of lean meat, backfat depth); pay rate (per kg dead weight); or if according to backfat depth, variation in P2, definition of maximum P2 for each grade class, pay rate for each grade class.

Output reports

An output report can be generated at any time or any weight during the growing phases from start to finish. The report represents the simulated response prediction relating to the input variables selected. It is not an optimum production profile; it is simply a statement of the consequences of the input activity. A range of input variables will generate a range of responses, from which the sensitivities and cost benefits can be judged and the best course of action selected. The following reports are offered by the Edinburgh model.

Efficiency reports: Live weight at end of period (or at slaughter); time taken to reach end of period (or slaughter); average feed intake during period; total feed intake; total feed used; daily live weight gain; feed conversion ratio.

Carcass reports: Carcass weight; carcass yield (KO%); per cent lean in carcass; lean meat yield; backfat depth; percentage of pigs in each grade class.

Biological reports: Energy costs of maintenance, protein deposition, fat deposition and cold thermogenesis; rates of protein and fat growth; total fatty tissues, lean tissues and bone in the carcass sides.

Financial reports: Number of batches of pigs through finishing house; total pigs sold per year; average value per kilogram dead weight; average value of live pigs at sale; total feed usages; feed costs per pig; margin per pig over feed and weaner costs; gross margin per pig; annual gross margin for the whole unit; annual net margin for the whole unit.

Case histories

Case 1 An attractive flat-rate contract for 70-kg pigs was offered to a producer already doing well with a quality premium scheme for 90-kg pigs. He fed a 17.5 per cent CP diet to a restricted scale and the farm margin was £7 000. The model was set up to operate close to his existing situation and then run to simulate the offered contract for a lower slaughter weight (Fig. 6.11). The lighter weight was not attractive unless and until the pigs were fed generously, while the best solution for slaughter at either weight also included use of two diets (see Fig. 6.11). In the event, all pigs were moved on to a more generous feeding scale and a two-diet feeding scheme, while half (mostly the castrates) were drawn out for sale at 70 kg and the remainder kept on until 90 kg (profit being further enhanced by an improvement in grading following withdrawal of the castrates at an earlier stage).

Figure 6.11 ***Simulations to examine the comparative advantages for one particular producer of slaughter at 70 or 90 kg.***

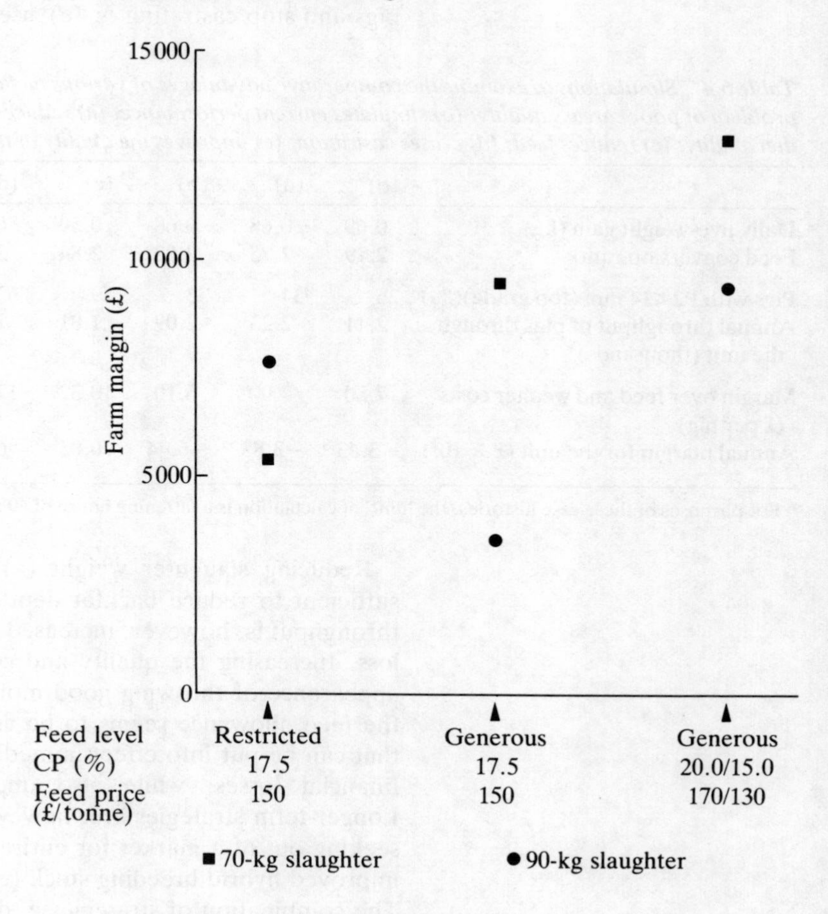

Case 2 A producer was alerted by his meat processor that his pigs would not be acceptable any longer unless something was done to make them less fat. Upon investigation of the accounts, he also discovered the unit was running at a loss. He was feeding mixed female and castrated commercial-type pigs to a ration scale using two diets (20–50 kg, 175 g CP per kilogram; 50–87 kg, 150 g CP per kilogram). Table 6.4 shows at (o) the simulation of his current performance at that time. Grading could be improved in five obvious ways: (a) reduce slaughter weight, (b) increase protein

content of the diets by 25 g/kg and pay an extra £10 per tonne, (c) reduce daily feed allowance, (d) find a market for entire male pigs and stop castrating or (e) use an improved strain of pig.

Table 6.4 Simulations to examine the comparative advantages of various tactics and strategies to resolve a problem of poor carcass quality: (o) simulates current performance; (a) reduces slaughter weight; (b) increases diet quality; (c) reduces feed; (d) ceases castration; (e) improves the quality of the pig

	(o)	(a)	(b)	(c)	(d)	(e)	(cde)	(bde)
Daily live-weight gain (kg)	0.69	0.68	0.68	0.59	0.71	0.70	0.60	0.73
Feed conversion ratio	2.79	2.75	2.82	2.84	2.69	2.74	2.78	2.57
Pigs with P2<14 mm (top grade)(%)	26	34	33	59	62	43	86	91
Annual throughput of pigs through the unit (thousands)*	2.11	2.23	2.09	1.81	2.15	2.13	1.84	2.23
Margin over feed and weaner costs (£ per pig)	7.50	7.00	6.10	10.32	12.03	9.84	12.75	13.24
Annual margin for the unit (£ × 10^3)	−3.43	−3.83	−6.44	0.05	6.54	1.66	4.78	10.12

* For purposes of these case histories, the 'unit' of calculation is a fattening house of 40 pens each holding 15 pigs.

Reducing slaughter weight (a) is no help because it is not sufficient to reduce backfat depths by any appreciable amount; throughput is, however, increased, resulting in more pigs sold at a loss. Increasing the quality and cost of the diets (b) has all the appearance of throwing good money after bad, but (c) reducing the feed allowance seems to be an appropriate short-term tactic that can be put into effect immediately and will stem the flow of financial losses, while also improving grading substantially. Longer-term strategies are, however, urgently required, with the seeking out of a market for entire males (d) and the purchase of improved hybrid breeding stock (e) being self-evidently essential. The combination of strategies c, d and e looked attractive to the producer, but simulation (cde) indicated that this was not likely to be so. The feed scale appropriate to the original stock was now too restrictive for improved strains; and, most importantly of all, the higher class of animal now required diets of improved quality, and tactic (b), to use better but more expensive diets, is now a vital part of the winning formula, which is (bde). At the point of having settled on a future long-term strategy, it was worth returning to the immediate possibility of combining the short-term tactic (c) of reducing the feeding level with the use of the grower diet (175 g

CP per kilogram) through to slaughter. This latter brought about an immediate improvement in carcass quality to 72 per cent top grade while the margin remained just positive. This at least amply satisfied the meat processor until such time as the longer-term strategies could be fully implemented.

Case 3 A past pedigree breeder, now operating as a multiplier for a breeding company, supplied stock to a second commercial unit also under his ownership. The meat processor consistently praised the high quality of the pigs received. Good-quality diets were rationed to mixed entire and female stock. The commercial unit was in profit (Table 6.5, (o)) but was not really earning its keep. The simple expedients of increasing the feed scale and slaughtering some 3 kg heavier (a) were able to double the profits without materially harming the good grading performance, which was important to sales from the parent multiplier herd.

Table 6.5 Simulations to examine the possibility of improving the profit of a high-quality herd requiring to maintain an excellent carcass quality record: (o) simulates the current position; (a) increases the feed allowance and slaughters at a slightly higher weight

	(o)	(a)
Daily live-weight gain (kg)	0.65	0.72
Feed conversion ratio	2.52	2.49
Pigs with P2<14 mm (%)	98	91
Annual throughput of pigs through unit (thousands)*	1.99	2.11
Margin over feed and weaner costs (£ per pig)	14.03	15.58
Annual margin for the unit (£ × 10^3)	3.95	8.60

* For purposes of these case histories, the 'unit' of calculation is a fattening house of 40 pens each holding 15 pigs.

Case 4 A producer, feeding his pigs *ad libitum* from self-feed hoppers was offered four alternative diets to be given from start to finish (Table 6.6). For this particular case, feed 3 was the best option (although this was not so for his neighbour, whose conditions of production and marketing were different). However, although 3 was the preferred diet, the pigs were clearly eating more than was good for carcass quality and grading. On the unit it was ascertained that feed intake could be restricted if the slides on the hoppers in the finishing house were closed down to reduce the rate of flow. The simulated consequences of such action was illuminating in that both diets 3 and 4 now performed similarly well, with growth rates of about 0.78 kg/day, just over 80 per cent of pigs in top grade and margins of around £10 000. The choice between diets 3 and 4 had thus become relatively insensitive.

Table 6.6 Simulation to examine the comparative advantages of four feeds of differing quality and price

	Feeds			
	1	2	3	4
Crude protein (g/kg)	150	175	200	225
Crude fibre (g/kg)	60	50	40	30
Oil (g/kg)	30	45	55	60
Price	140	155	165	175
	Simulated responses			
Daily live-weight gain (kg)	0.68	0.78	0.84	0.86
Feed conversion ratio	2.97	2.62	2.49	2.42
Pigs in top grade (%)	26	50	55	57
Annual throughput (thousands)	2.11	2.38	2.52	2.61
Margin over feed and weaner costs (£ per pig)	6.21	9.39	10.14	9.00
Annual margin for the unit (£ × 10^3)	−6.13	2.60	5.49	3.24

Comparing sensitivities is a vital part of the effective use of simulation models, and these will change dramatically with fluctuations in the average price for pigs, the average price for pig feeds, and the premiums paid for quality with respect to both diet inputs and carcass outputs. The above cases related to carcass grade premiums of around £0.10 per kilogram dead weight. Greater or lesser grade differentials would have resulted in quite different conclusions to the various simulations.

Summary

- Response prediction is a required part of good management.
- The computer revolution allows not only the construction of simulation models, but also the electronic interconnexion of data collection with response prediction.
- Deductive models are more complex to construct than empirical models, but are more likely to give valid predictions.

The need for hypotheses is basic to model building activity and necessitates a view of science that goes beyond the mere restatement of applied feeding trials.

- To be effective management aids, models work best when the input variables reflect the control points of the system about which information is readily available.
- The construction of a model requires a core that incorporates the current state of the art of the scientific understanding of biology, together with a substantial framework of financial information and systems analysis.
- It is an explicit part of the philosophy of use of simulation models for response prediction that each simulation is unique and may not be used to generalize. Changes in both general and particular circumstances have dramatic effects on the outcome of 'what if' questions and on the relative sensitivities of the various production parameters. Response prediction models must therefore: (a) be set up specific to place, time and circumstance, and (b) be accessible for frequent interrogation.

Index